T0371970

Science, Policy and Stakeholders in Water Management

An Integrated Approach to River Basin Management

Edited by
Geoffrey D. Gooch and Per Stålnacke

London • Washington, DC

First published in 2010 by Earthscan

Earthscan Ltd, Dunstan House, 14a St Cross Street, London EC1N 8XA, UK
Earthscan LLC, 1616 P Street, NW, Washington, DC 20036, USA
Earthscan publishes in association with the International Institute for Environment and Development

For more information on Earthscan publications, see www.earthscan.co.uk or write to earthinfo@earthscan.co.uk

ISBN: 978-1-84407-919-3 hardback

Typeset by JS Typesetting Ltd, Porthcawl, Mid Glamorgan
Cover design by Dan Bramall

A catalogue record for this book is available from the British Library

Library of Congress Cataloging-in-Publication Data

Science, policy, and stakeholders in water management : an integrated approach to river basin management / edited by Geoffrey Gooch and Per St?lnacke.
p. cm.
Includes bibliographical references and index.
ISBN 978-1-84407-919-3 (hardback)
1. Watershed management–Europe–Citizen participation. 2. Watershed management–India–Citizen participation. 3. Watershed management–Southeast Asia–Citizen participation. I. Gooch, Geoffrey D. II. St?lnacke, Per.
TC455.535 2010
363.6'1094–dc22

2009052604

At Earthscan we strive to minimize our environmental impacts and carbon footprint through reducing waste, recycling and offsetting our CO_2 emissions, including those created through publication of this book. For more details of our environmental policy, see www.earthscan.co.uk.

Printed and bound in the UK by TJ International.
The paper used is FSC certified and the inks are vegetable based.

Contents

List of Figures and Tables

Figures

Tables

List of Contributors

Andrew Allan (Master of Laws) is a lecturer at the United Nations Educational, Scientific and Cultural Organization (UNESCO) Centre for Water Law, Policy and Science, University of Dundee, Scotland. His research interests concern national water allocation frameworks and governance of water resources, and include implementation of integrated water resource management (IWRM), flood management, participatory irrigation systems and the effectiveness of governance regimes. Most recently, he has provided expert advice to the government of Kazakhstan on improving the law affecting farmer-managed irrigation systems, and has developed a system of indicators to evaluate the implementation of governance in the context of climate change as part of the European Commission (EC) project BRAHMATWINN.

António Betamio de Almeida is an emeritus professor at the Technical University of Lisbon, Portugal, and a retired full professor of the Civil Engineering Department (Hydraulics and Water Resources Division) of the Technical Superior Institute (IST) at Lisbon, Portugal. He is an expert in risk management and analysis and in hydrodynamics. He is currently a consulting engineer (hydraulics systems), a member of the Centro de Estudos de Hidrossistemas (CEHIDRO) research centre and an elected member of the National Academy of Engineering of Portugal.

Susan Baggett is a researcher at the UNESCO Centre for Water Law, Policy and Science, University of Dundee, Scotland. With a PhD in water management from Cranfield University and an MRes in environmental science from University College London, her research interests include stakeholder participation and interaction. She was a member of the European Union (EU) AQUAREC project, and has worked as a research analyst in the private sector and as an environmental strategy officer, focusing primarily on climate change, in the public sector.

Line J. Barkved (MSc hydrology) has been employed at the Norwegian Institute for Water Research (NIVA) since 2001, where she currently holds a permanent position as research scientist. Her work is mainly related to modelling of hydrology, nutrient losses and climate change. She has working experience from the Arctic, Europe, India and Vietnam. During her career, she has had a central role in the coordination team of several EU projects, including the STRIVER project.

David N. Barton (senior scientist at NIVA) holds a PhD in agricultural and resource economics from the Agricultural University of Norway (NLH). Barton has eight years' experience in the use of environmental and resource economics in water resource management, sustainable harvesting of wetland resources and conservation management planning. He has specialized in the application of non-market valuation methods for environmental quality, particularly water and sanitation services, as well as an evaluation of institutional compensation mechanisms. He has applied economic valuation methods and extended benefit-cost analysis to the evaluation of multiple-use conflict in wetlands management.

Dag Berge is senior research scientist at NIVA. Originally educated as a limnologist, his main expertise is the eutrophication of lakes and reservoirs, phosphorus loading models and phytoplankton. During the last 15 years he has been working on an environmental impact assessment (EIA) of hydropower development in Asia and Latin America, as well as water management planning, pollution abatement planning, monitoring of lakes and rivers, and drinking water supply. He has project experience spanning more than 20 countries.

Fayçal Bouraoui holds a permanent position as senior researcher at the Joint Research Centre of the European Commission in the Institute for Environment and Sustainability. He is the leader of the FATE research group. His research involves the development and application of modelling tools at continental scale to evaluate the effects of current European policies on water quality and to assess the effects of possible future scenarios of climate and land-use changes on water resources. His research interests include the impacts of agricultural activities upon water quality in Europe and in Africa, under the challenge of global changes.

Dale Campbell has been a research associate for the STRIVER project at the UNESCO Centre for Water Law, Policy and Science at the University of Dundee, Scotland. With an MSc in environmental management from the University of London and 20 years of experience in environmental and development projects, she has worked on issues related to river basin management in Asia, Europe and Canada. Her research interests include procedural rules related to river basin commissions and the implementation of international conventions such as the United Nations Framework Convention on Climate Change (UNFCCC), the United Nations Convention to Combat Desertification (UNCCD) and the Convention on Biological Diversity (CBD), as well as the multilateral United Nations Economic Commission for Europe (UNECE) conventions.

Johannes Deelstra holds a permanent position as senior researcher at Bioforsk, the Norwegian Institute for Agricultural and Environmental Research. Deelstra studied irrigation and drainage at the Agricultural University of Wageningen, The Netherlands. He has considerable experience in issues related to the interaction between agriculture and the environment, and has a special interest in the role that hydrology plays in nutrient transport and erosion in small agricultural catchments. He has worked in Norway, the Baltic countries, Russia, Central and South America and Africa.

Geoffrey D. Gooch is a professor of political science at Linköpings University, Sweden, and at the UNESCO Centre for Water Law, Policy and Science at the University of Dundee, Scotland. He is an expert in water and environmental policy analysis, public participation and communication, and he has conducted research in many parts of the world. He is currently coordinating the EU funded 'LiveDiverse' project which studies biodiversity and livelihoods in Costa Rica, Vietnam, South Africa and India.

Bruna Grizzetti is a researcher at the Joint Research Centre of the European Commission in the Institute for Environment and Sustainability. She has experience in hydrological and nutrient modelling at a continental scale in support of the development and implementation of environmental European policies. Her research interests involve interdisciplinary projects for sustainable water management in developing countries. Since 2007, Bruna Grizzetti has been a member of the Coordination Team of the European Nitrogen Assessment, funded by the European Science Foundation.

K. J. Joy is a senior fellow with the Society for Promoting Participative Ecosystem Management (SOPPECOM), Pune, India. His areas of work and interests include people's institutions for natural resource management, drought and drought-proofing, participatory irrigation management, river basin management and multi-stakeholder processes, watershed-based development, water conflicts, and people's movements. He is currently coordinating the Forum for Policy Dialogue on Water Conflicts in India and is also involved in the EU-supported LiveDiverse project, which studies biodiversity, vulnerability and livelihoods.

Marta Machado is a biophysical engineer and an expert in geographic information systems (GIS). She currently works in a private company as a consultant and GIS supervisor in environmental impact assessment studies.

S. Manasi is currently working as assistant professor in the Centre for Ecological Economics and Natural Resources, Institute for Social and Economic Change, Bangalore, India. She has worked in various areas of water resources – rural water supplies, urban water supplies and lately on integrated water resources management. Her areas of research are water resources, solid waste management and climate change.

N. Latha is a senior research assistant at the Centre for Ecological Economics and Natural Resources, Institute for Social and Economic Change, Bangalore, India. Her areas of research interest are water resources, wastewater management and climate change.

Udaya Sekhar Nagothu is a research scientist at Bioforsk and holds a PhD in development studies from the Agriculture University of Norway. He has considerable research experience in institutional and policy analysis, stakeholder analysis and stakeholder participation, conflict analysis and conflict management, tenure and property rights, and integrated watershed management. He recently participated in a project on triangular institutional cooperation programme (Ethiopia–India–Norway), and has worked in Vietnam, India and eastern Sri Lanka.

Ingrid Nesheim is a researcher at the Centre for Development and the Environment, University of Oslo, Norway. Nesheim is qualified in the field of plant ecology, ethnobotany and multivariate statistics. She has long experience with interdisciplinary research in various research projects involving fieldwork from rural areas in the tropics, and her perspective encompasses both environmental and social science. She teaches interdisciplinary methodology at the Masters level at the University of Oslo, Norway.

Dang Thi Kim Nhung is a researcher at the Institute of Geography, Vietnam Academy of Science and Technology, and a geographer with a PhD in geography from the IoG Poland Scientific Academy. She has wide experience in EIAs and environmental management, and has participated in many international projects funded by the Association of Southeast Asian Nations (ASEAN) and the EU.

Suhas Paranjape is senior fellow with the Society for Promoting Participative Ecosystem Management, Pune, India. His areas of work and interests include participative natural resource management, renewable energy and materials, drought and drought-proofing, participatory irrigation management, people's science movements and the history of science, regional planning, watershed-based development, and water conflicts and social movements. He is currently involved in the Forum for Policy Dialogue on Water Conflicts in India and also in the EU supported LiveDiverse project, which studies biodiversity, vulnerability and livelihoods.

Maria Manuela Portela is a civil engineer. She has a PhD degree in civil engineering and an MSc in hydraulic and water resources. She is an assistant professor of the Environment and Water Resources Division of the Civil Engineering Department of the Technical Superior Institute (IST), Lisbon, Portugal, as well as a member of the Centro de Estudos de Hidrossistemas (CEHIDRO) Research Centre. She is an expert in hydrology, hydrologic modelling (conceptual and statistical models), water resources management and small hydropower scheme design. In addition to university research and teaching, she continues to develop consulting activity in private companies.

Dr Antonio Lo Porto is a researcher at the Italian Water Research Institute (IRSA-CNR). His area of interest is water resources management at watershed scale with particular attention to arid and semi-arid areas and to agricultural impacts on water. He has been involved in several research projects in Europe, Northern Africa, South America and India, and is a member of the Management Committee of the COST Action 869 on 'Mitigation Options for Nutrient Reductions in Surface Waters and Groundwaters'. He teaches 'Cartography and Landscape Planning' at the Tuscia University in Cittaducale.

Santiago Beguería Portugues is a post-doctoral fellow and has a PhD in geography from the University of Zaragoza, Spain. He has published in the main international journals in hydrology. His primary works are related to rainfall–runoff relationships, extreme events and the location of sediment sources. He is an expert in GIS and remote sensing.

K. V. Raju is a professor at the Institute for Social and Economic Change, Bangalore, India. Raju has long-term experience in this field and is currently working as the economic adviser to the chief minister of Karnataka State. He specializes in social, environmental and institutional studies of water management and rural development.

Alistair Rieu-Clarke (LLB, LLM, PhD) is a senior lecturer at the International Hydrological Programme (IHP) HELP Centre for Water Law, Policy and Science, University of Dundee, Scotland. He is actively involved in a range of multidisciplinary international research collaborations largely centring on measuring the effectiveness of (transboundary water) governance regimes and designing participatory research methods in the context of water resources management. Additionally, he runs a postgraduate module on the international law of water resources as part of the UNESCO Centre's Water Law: Water Leaders Programme, and supervises a number of PhDs.

Per Stålnacke holds a permanent position as senior research scientist and is head of Water Quality and Hydrology at Bioforsk, the Norwegian Institute for Agricultural and Environmental Research. Stålnacke has long-term experience in issues devoted to integrated water resources management, with particular emphasis on studies of pollutant fluxes in river basins and statistical analysis of historical environmental monitoring data. Stålnacke has developed statistical models for source apportionment of pollutants and time-trend analysis. He has considerable experience in the Baltic Sea region, particularly in the Baltic States, Poland and Russia; since 2006, he has also worked in India, Vietnam and Cambodia.

Preface

During recent years we have seen an evolution away from the more 'traditional', typically sector-oriented (water supply, irrigation, hydropower, industrial plants) style of water management that focused on satisfying perceived demands within each sector, towards 'integrated' management attempts to take a cross-sectorial approach, with a focus on the adequate management of all waters, and on the demand, supply and use of water. Despite these recent changes in management structures and conceptual thinking, there is, however, still a lack of 'success stories', of examples of how challenges to traditional water management have been met, and how solutions have been developed to the problems of an integrated approach.

One important prerequisite to successfully solving these challenges is the ability to involve stakeholders (especially the local ones 'using' the water for day-to-day activities and livelihoods). There is also a need for developing ways to improve transparency at all administrative and sectorial levels, including policy-making and its implementation. In this book we claim that the interaction of different forms of knowledge and the usage and uptake of scientific results are crucial in this respect.

More specifically, this book examines one of the major problems facing practitioners and scientists working with water management – how to integrate knowledge and experiences from the scientific, policy and stakeholder perspectives. This science–policy–stakeholder interface (SPSI) is examined in the book both analytically and through the description of practical experiences from river basins in Europe, India and South-East Asia. This combination of theoretical and empirical work is unusual in the field of water management and integrated water resource management (IWRM) and will hopefully contribute to the development of the SPSI for practical policy purposes.

Following Chapters 1 and 2, which lay the framework for the book and discuss experiences from other projects, the authors go on to describe how SPSI was managed in the four case basins and how stakeholder participation (Chapter 3) and scenarios (Chapter 4) were used to integrate different perspectives, and to facilitate the communication of different forms of knowledge.

In Chapters 5 to 8, four important aspects of water management and SPSI are treated: these are water pollution (Chapter 5), land and water interaction (Chapter 6), environmental flow (Chapter 7) and transboundary water regimes (Chapter 8). The final chapter (Chapter 9) in the book analyses the SPSI context in water management in the four basins and provides a series of generic recommendations for improving the science–policy–stakeholder interface in water management.

This book stems from the STRIVER research project (www.striver.no). STRIVER stands for Strategy and Methodology for Improved IWRM – An Integrated Interdisciplinary Assessment in Four Twinning River Basins, and was a three-year European Commission (EC)-funded project (2006 to 2009) under the Sixth Framework Programme (FP6), contract no 037141.

The project comprised 13 research institutes from nine countries. In total, about 40 scientists and students (MScs and PhDs) with different backgrounds, both thematically and culturally, worked on the project. The majority of them have also contributed directly to this book in the following chapters; others have contributed indirectly through their research. This melting pot of scientific knowledge created a unique possibility to explore integrated water resource management and SPSI from various angles and entry points. Behind the chapters presented here lie many hours of discussions within the research group; but equally important are the discussions with stakeholders in the basins. These stakeholders included a wide array of interests and expertise, from local fishermen and water managers to policy-makers.

We would like to thank everyone who has contributed to the project and the book. All of the chapter authors and others who contributed in one way or another are acknowledged. Thanks also to the advisory board for the STRIVER project which had representatives from international water organizations, such as the Global Water Partnership, the United Nations Economic Commission for Europe (UNECE), the United Nations Educational, Scientific and Cultural Organization (UNESCO), the International Network of Basin Organizations (INBO), river basin commissions, and water-user associations, such as the Mekong River Commission and the Glomma River Basin Water Management Association (GLB), Norway. A special thanks to all the stakeholders and managers in the four case basins – Glomma (Norway), Tagus (Spain and Portugal), Tungabhadra (India) and Sesan (Vietnam and Cambodia) – who have interacted with us in a very fruitful way during field visits and stakeholder meetings. This interaction was a great source of inspiration when writing the book. Last, but not least, a very big thanks to Dr Sue Baggett, who in an excellent fashion has acted as a secretarial editor for this book volume and who also contributed to two of the chapters.

Geoffrey D. Gooch and Per Stålnacke
26 January 2010

List of Acronyms and Abbreviations

ADB	Asian Development Bank
ANT	actor network theory
ASEAN	Association of Southeast Asian Nations
CADA	Command Area Development Authority
CBD	Convention on Biological Diversity
CEHIDRO	Centro de Estudos de Hidrossistemas (Portugal)
CIS	Common Implementation Strategy (*of the* WFD)
CSO	civil society organization
DPR	detailed project report
DSS	Decision Support System
EC	European Commission
EF	environmental flows
EIA	environmental impact assessment
ENGO	environmental non-governmental organization
EU	European Union
EU SDS	European Strategy for Sustainable Development
EVN	Electricity of Vietnam
FP6	Sixth Framework Programme
FSL	full supply level
GIS	geographic information system
GLB	Glommens og Laagens Brukseierforening (Glomma River Basin Water Management Association)
GWh	gigawatt hours
GWP	Global Water Partnership
ha	hectares
HEP	hydroelectric power
HPP	hydropower project
HRA	hydrographic region administration
HRMP	Hydrographic Region Management Plan
IHP	International Hydrological Programme
INBO	International Network of Basin Organizations
IPCC	Intergovernmental Panel on Climate Change
IRSA–CNR	Water Research Institute–National Research Council (L'Istituto di Ricerca sulle Acque–Consiglio Nazionale delle Ricerche)
IST	Technical Superior Institute (Portugal)

IWMI	International Water Management Institute
IWRM	integrated water resource management
KSPCB	Karnataka State Pollution Control Board
KWDT	Krishna Water Dispute Tribunal
LEISA	low external-input sustainable agriculture
m	metres
Mm^3	cubic megametre
MARD	Ministry of Agriculture and Rural Development
MCA	multiple-criteria analysis
MEA	Millennium Ecosystem Assessment
MOFA	Ministry of Foreign Affairs
MOL	minimum operation level
MONRE	Ministry of Natural Resources and Environment
MOWRAM	Ministry of Water Resources and Management (Cambodia)
MPI	Ministry of Planning and Investment
MRC	Mekong River Commission
MSP	multiple stakeholder platform
N	nitrogen
NDVI	Normalized Difference Vegetation Index
NGO	non-governmental organization
NIVA	Norwegian Institute for Water Research
NLH	Agricultural University of Norway
NOAHH	National Oceanic and Atmospheric Administration
NVE	Norwegian Water and Energy Directorate
OED	Ministry of Petroleum and Energy (Norway)
P	phosphorus
PBL	Plan and Building Law (Norway)
PI	Pressure–Impact
PIM	participatory irrigation management
PIMCEFA	Pressure Impact Multi-Criteria Environmental Flow Analysis
PP	public participation
PTA	participatory technology assessment
RBMP	river basin management plan
R&D	research and development
SADC	Southern African Development Community
SHG	self-help group
SIA	social impact assessment
SOPPECOM	Society for Promoting Participative Ecosystem Management
SPSI	science–policy–stakeholder interface
SRI	System of Rice Intensification
STRIVER	Strategy and Methodology for Improved IWRM – An Integrated Interdisciplinary Assessment in Four Twinning River Basins
TBSB	Tungabhadra Sub-Basin

UK	United Kingdom
UNCCD	United Nations Convention to Combat Desertification
UNCED	United Nations Conference on Environment and Development
UNDP	United Nations Development Programme
UNECE	United Nations Economic Commission for Europe
UNESCAP	United Nations Economic and Social Commission for Asia and the Pacific
UNESCO	United Nations Educational, Scientific and Cultural Organization
UNFCCC	United Nations Framework Convention on Climate Change
US	United States
WALMI	Water and Land Management Institute
WCED	World Commission on Environment and Development
WERF	Water Environment Research Foundation
WFD	Water Framework Directive
WRM	water resource management
WUA	Water User Association

1

Introduction: The Science–Policy–Stakeholder Interface (SPSI)

Geoffrey D. Gooch and Per Stålnacke

Introduction

Sustainable water management is now the focus of concern for many different groups in society, including scientists, politicians, water managers, the public, non-governmental organizations (NGOs) and industrialists. Their concerns are diverse, however, ranging from the effects of increasing demands on the quantity and economic uses of water, to the environmental quality of water and aquatic life. In addition to worries about the effects of global change on the world's freshwater resources (Bates et al, 2008), there are concerns about the problems created by expected sea-level rises reflected through climate change projections (Jenkins et al, 2009), over-extraction of water from underground aquifers and increasing demands from growing populations (Vörösmarty et al, 2000). These have all contributed to an intensified interest in water among different sectors. However, while each of these sectors attempt to understand and manage water issues according to their own rationality and comprehension of the world, in many cases they seem to have difficulties understanding the rationalities of the other groups. Therefore, while much effort is being put into improving water management and water ecology internally within these sectors, it appears difficult to create a combined integrated impetus for sustainable water use.

Integrated water resource management (IWRM) has been proposed as a way out of this predicament; yet IWRM, as such, does not necessarily target the difficulties outlined above. IWRM provides a general normative framework that specifies an integrated approach, and this is in itself a major step forward from the traditional sector-specific water paradigm. However, what is needed is a deeper understanding of exactly why this integration often proves so difficult to achieve, as well as examples of how different groups might be helped to work closer together.

In this book we attempt to do just that, to analyse what integration in water management is about, to unravel some of the main obstacles to this integration, and also to provide some good examples from different parts of the world on how integration can be improved. In this first chapter we look at the three main 'groups of groups' – namely, the science, policy and stakeholder communities who are then examined in the following chapters, including the conglomerates of interests that these main groups of actors represent. These are then examined in more detail in the individual chapters, where it will be shown that, while it is convenient to present these groups on first examination as homogeneous, they represent combinations of widely differing interests and bring varying perspectives to the table. For example, while we look at 'science' as a group in this initial analysis, we will then go on to show in Chapter 2 that 'science' consists of a wide variety of competing interests, rationalities and forms of explanation. The rationality of economics does not necessarily combine comfortably with that of ecology, nor does the understanding of the world presented by anthropologists fit easily into the mathematical constructions of the modellers. Yet, all of these disciplines are parts of 'science' and all of them have something to contribute to improving water management. In a similar way we discuss 'stakeholders' in this introductory chapter as one of the major groups, and then in Chapter 3 we disaggregate the group and show that these 'stakeholders' represent many different, and often competing, interests. We also begin here by talking about 'policy' as a central aspect, to then go on in Chapters 8 and 9 to analyse just how diverse policy communities can be, as can be their aims and preferred methods of achieving those aims.

In Chapters 3 and 4 we describe some of the methods that have been used by the authors working in Europe, India and South-East Asia. These have been utilized to facilitate interaction between the main groups described in this chapter, as well as in Chapters 2, 3 and 8. These methods include stakeholder workshops, focus groups, interviews and scenario-building. What is specific about the way in which they have been used in the cases described in this book is that a conscious focus has been on the exchange of knowledge between groups, as well as on furthering understanding of the different groups' rationalities. The authors have used these methods, together with others, to try to develop ways of improving the science–policy–stakeholder interface (SPSI) that is the central issue of this book.

In Chapters 5 to 8 the interaction of the groups in specific areas of water management are explored. These areas are water pollution, land and water use, environmental flow and transboundary water regimes. In Chapter 9 the implications of the interactions in the four case basins are discussed, and in this final chapter we look at how the experiences from the four case basins, and from the water issues discussed in Chapters 5 to 8, can contribute to improving the interaction of the different groups that we have identified as central to water management. Let us for now, however, turn our attention to the groups that form the focus of this book. These are the scientific community, policy-makers, managers and stakeholders. Each will be discussed in turn; but our focus is not only on the internal divisions within them, but also on the interaction between them: on the SPSI.

A well-known dilemma in water management is the incorporation of (or lack of) scientific information within the policy and management processes. This science–policy interface seems to be the source of much frustration and misunderstanding (Roux et al, 2006). Scientists, whether from the natural or social sciences, complain that policy-makers and managers do not take into account the knowledge generated within their disciplines. Conversely, from the other side, the complaint is that scientists do not provide the kind of knowledge necessary to solve day-to-day problems. A significant factor here seems to be connected with the ways in which problems are formulated, as scientists tend to focus on problems that can be solved, or at least analysed, through the methods and methodologies that they are trained to use. They are also concerned with problem-solving procedures that fit into the reporting systems of their scientific disciplines, which can be judged (preferably positively) by others in their scientific community. Policy-makers and managers, on the other hand, usually need practical answers to immediate problems. Policy-makers, in their role as politicians or elected officials, also need to take into account the perceptions of their electorate, the people who hopefully will vote for them again the next time around. The differences between these two ways of formulating and understanding problems is significant and sometimes leads to claims that one side lacks a genuine interest in understanding the other's perspectives. This is not necessarily so. A frequently proposed solution (from scientists) is that policy-makers must be *made* to understand their results, while for policy-makers the answer is often that scientists should be more sensitive to policy and management needs.

The science–policy interface is attracting a growing amount of interest; yet it is only part of the problem. The growth of interest in, and the demands made for, increased stakeholder and public participation provide another potential interaction: the science–stakeholder and the policy–stakeholder interface. For the sake of simplicity, let us envisage these as a triangle or pyramid with three apexes – namely, science, policy and stakeholders, although not necessarily in that order. Gooch coined the term 'trialogue' in an application submitted to the European Union (EU) in 2003 and in a paper at the Stockholm Water Week in 2004 (Gooch, 2004a) to describe this basically normative interaction of science–society–policy. The concept has since been used by other authors, such as Turton (2008). However, a 'trialogue', like a 'dialogue', infers a functioning communication process between the different parts, and the basis of the problem approached in this book is that there is not much evidence of this happening yet. In this instance, in order to stress that it is the interaction that we consider more important, we will therefore talk about the 'triple interface' rather than a 'trialogue'. This leaves the more theoretical and normative aspects of trialogue to one side in order to focus on the communication mechanisms between groups and the methods which can facilitate this communication and interaction.

How, then, do you include science, policy and stakeholders in management, decision-making, policy-making and implementation in such a technically complicated field as sustainable water management? Are stakeholders and the public able to understand the complexity of these issues? Can scientists break down their disciplinary barriers and work together? Can politicians look ahead, beyond the horizon of the next round of elections four to five years in the future? The answer to these questions is yes: we believe that they can and should, under certain conditions. We believe that it is possible for these groups, including stakeholders and the public, to both understand and contribute to policy for sustainable water management. A wide range of tools are now available that enable different groups to participate in environmental management (Gooch and Huitema, 2007), such as round tables, citizen juries, panels, Delphi processes, etc., and we will discuss these in more detail in Chapter 3. However, while many of these methods can help laymen to understand complex water issues, they may not provide a means of involving non-(water) experts in water management. We deliberately stress that just as stakeholders and the public have forms of knowledge vital to sustainable water management, scientists and managers who are also not specifically expert in water management can also contribute to the field through their understanding of many of the issues that are central to water issues. Water management is not a standalone issue; it is embedded in all other societal issues.

What is science in the science–policy–stakeholder interface?

Before considering the interface between the three basic groups identified in this book as central to water management, we need to look, in turn, at what we mean by 'science', 'policy' and 'stakeholders'.

First, let us look at the role of the scientists and experts who provide information, models and analyses on water issues. Here we can see that by 'scientists' we do not simply mean representatives for the natural sciences (biology, physics, chemistry, etc.), but also experts in hydrology, economics, law and the social sciences, etc. These scientists and experts can contribute with natural scientific information, managerial expertise, policy analyses and other forms of scientific and academic knowledge. What they have in common is that they all claim to constitute some form of objective knowledge and they are all somewhat outside of the policy and management systems, even if they act as advisers and formulators of proposals. They do not take decisions on policy, they are not directly responsible for the consequences of their proposals (except as academics), and they are not dependent on being voted in or out of office. This special role of science and experts in water management calls for particular attention. While many of them claim to be quite objective in their judgements and recommendations, expert opinions are far from value free. Scientists and experts occupy their own worlds of systems of norms and values, or paradigms (Kuhn, 1970).

Knowledge for water management can and should, of course, come from a number of different disciplines. However, as noted, each individual discipline has its own set of rules and criteria for success. So before we can look at the interaction between our three main groups of actors in water management, we need to look at the interaction within the different groups of scientists and experts. More will be said about this in Chapter 2; but we also need to stress that not only must the knowledge and expertise within the science group, as a group, be integrated, combining the natural and social sciences (Gooch and Stålnacke 2006), but that it is also necessary to improve the ways in which this knowledge is presented and communicated to other actors in water management (Gooch, 2004b; Gooch and Stålnacke, 2006), both within the science group and between the science and other groups. This involves an interaction between scientific knowledge and other forms of knowledge, and an acceptance by representatives of the different science groups that other forms of knowledge are also legitimate. In this combination of knowledge forms, natural scientists focus on the physical processes (Mostert, 1999) in water ecosystems, while the social sciences can contribute to the understanding of the structure, institutions, ideas and strategies of actors and the 'management' of the decision-making process (Klijn and Koppenjan, 1997). Both aspects are necessary components of the science group, together with many others.

Within our triple interface, science enjoys a special position. Science is not only employed by policy-makers and authorities, but also by the other groups. non-governmental organizations (NGOs), civil society organizations (CSOs), international organizations and business all utilize scientific results as a means of legitimizing their standpoints. Environmental groups use scientific knowledge to support their demands to protect the environment, just as business uses science to defend the suitability of its actions, and policy-makers and water managers claim that their policies are based on 'sound science'. These groups, however, do not necessarily obtain their support from the same sub-group of science, or even the same proponents of different scientific explanations. As noted, the science group is far from homogeneous, and within it there can be found support for widely different points of view, ranging from one view that water systems are extremely vulnerable to the other extreme that the same systems are resilient. This can result in competing scientific and expert opinions being used in disputes over water management. A key issue to consider is, therefore, what systems are in place to reconcile competing claims?

What is policy in the science–policy–stakeholder interface?

Policy is a many-headed beast, the result of competing perspectives and visions of the future, based partly on what is desirable and partly on what is possible. Policy is, to put it very simply, a designated way ahead from where we are now to where we would like to be in the future, with (hopefully) a road map of how to get there. While this definition is, of course, an extreme simplification, it does capture some of the vital elements of policy, which are that there is an idea of what we want and an idea of the best way to get it. However, as we all know, getting from A to B can be a complicated process, especially if road works or floods make the road difficult, or if we run out of fuel. This is why policy cannot be seen as a one-time process, as something that once decided must be adhered to. For the purposes of this book this also means that we see policy not only as the prerogative of the *policy-makers*, but also as a result of the influence of the *policy-doers*, the water managers and bureaucrats whose job it is to make policy happen. This is termed the implementation process in policy literature and is considered an especially problematic part of the policy process, as the intentions of policy-makers are often difficult to translate into actions.

Concerning the relationship between policy and science, while the polity (politics and administration) has, to a large extent, to rely on information from the scientific community, it must also work in a political context where one of the main issues is 'who gets what, when and how'. Policy-makers need to decide on how limited resources should be distributed. How should state income be used – to increase pensions for the elderly or to improve water quality in this or that river? Policy-makers must take into account the pros and cons of different choices of policy within their governance systems (which include law; see Chapter 8) as well as the information provided by the scientific community.

Besides the distinction between policy-makers and policy-doers outlined above, the policy group also consists of much more than representatives for political and administrative systems. Economic and business interests also make policy. They formulate and implement decisions on where and what to invest in, which technical innovations to support, and which road ahead is most likely to result in the best profits. They also have to formulate policies that take into account the demands of the political and legal systems in which they work, otherwise they may be fined, as well as bearing in mind the preferences of their customers who purchase their products.

The policy group described in this book therefore consists of a diverse group of policy-makers and policy-doers from political, governmental and economic spheres. We therefore look at the type of knowledge and information that this group can contribute to water management, and the kind of knowledge and information that it needs from other groups. It is clear that the types of knowledge and rationalities of political and economic policy-makers are different, and that they are both different from the type of knowledge of the scientific community. While scientists make claims of objectivity, political policy-makers claim that they use information for the common good of their communities, while economic policy-makers use information and knowledge to make money.

Who are the stakeholders in the science–policy–stakeholder interface?

Our third general category of actors is stakeholders. As will be explained in more detail in Chapter 3, these can consist of a wide variety of different individuals or groups. They may be economic stakeholders; owners of irrigated land, hydroelectric power (HEP) stations or water rights; managers of water utilities outside of the policy process; or they may be NGOs working with water issues or fishermen. A traditional way of defining these groups would be to say that they have a vested interest in water in some way in that they might gain or lose from policies affecting the river, lake or water body. However, this is too narrow a definition as it excludes groups who either do not personally benefit or lose out due to a certain way of managing water. Environmental NGOs, for example, claim that their activities are not dependent on the amount of personal loss or gain at stake, but that they are the result of a concern borne from a much broader perspective that might include the 'Earth' as an entity and the rights of future generations.

What is specific about SPSI in water management?

Before answering this question we need to ask ourselves what the science–policy–stakeholder interface is, in general, if there is such a context as a general SPSI. The term 'interface' implies some kind of interaction, of entities coming together; but for our purposes in this book this is too broad, too unfocused. First, we must look instead at what it is that is being communicated. We have already mentioned information and knowledge and these are, of course, at the centre of our interest when examining water management. However, the next question is knowledge and information for what? Here we can refine our categorization and perhaps use the working hypothesis that in many cases we are talking about risks, about threats and about ways to try to manage these. As can be seen from our chapter topics, we examine in this book a number of problematic issues. We look at water pollution, at the problems of land and water use, at the challenges of securing environmental flow and at the complications resulting from transboundary management of water resources. All of these constitute problems and risks towards humans and the environment, risks that need to managed in some way.

The focus of the SPSI examined in this book is therefore the risks, threats and problems affecting water management, and possible ways to alleviate these. We claim that better communication between the different actors in water management is central to this ambition to solve problems, and we present examples of how the authors have attempted to improve SPSI in a number of issues central to water management. The final two chapters summarize the experiences gained from the case basins we have worked within for the last three years, and finally we draw conclusions for SPSI in water management that hopefully will be of use to others working in this vital field, whether they are part of the scientific, policy or stakeholder communities.

Our case basins

The work presented here is the result of studies of SPSI and water management in four case basins in different parts of the world: the Glomma River in Norway; the Tagus River that flows between Spain and Portugal; the Tungabhadra River in India; and the Sesan River that flows from Vietnam into Cambodia. These basins are described briefly below; more information about the basins can be found in Chapter 9. It will be shown in the individual chapters that while some issues, such as stakeholder participation, have been developed in all four basins, others have not. Water pollution issues (Chapter 5) were examined in the Glomma in Norway and in the Tungabhadra in India; land- and water-use interaction (Chapter 6) was studied in the Tagus (Spain and Portugal) and the Tungabhadra (India); environmental flow methodologies (Chapter 7) were developed in the Glomma (Norway) and in the Sesan (Vietnam and Cambodia). Finally, transboundary water regimes (Chapter 8) were studied in most detail in the Sesan (Vietnam and Cambodia) and the Tagus (Spain and Portugal). Our motivation for this was twofold: first, there was an ambition to facilitate an exchange of knowledge and experiences between the different basins, and one approach used to achieve this was twinning – the deliberate development of permanent or semi-permanent ties between researchers and practitioners in the basins. The second reason for the limitation of certain aspects to certain basins is tied to the first: resources were, as always, limited and it was necessary to focus on specific issues in specific basins in order to accomplish in-depth studies.

Because of the ambition to facilitate twinning we were faced with not only the interaction of actors in the science–policy–stakeholder interface in one basin, but also the interaction of these groups between two basins. As will be seen in the following chapters, the success of these twinning efforts varied between issues and basins. While the scientific twinning succeeded in most aspects, it was more difficult with the other groups of actors in the SPSI framework. We were also faced with the extra complication of transboundary rivers in two of our case areas: the Tagus (Spain and Portugal) and the Sesan (Vietnam and Cambodia). In a third basin, the Tungabhadra in India, while the river is not transboundary in the sense that it is transnational, it is trans-state and as such shares a considerable number of characteristics with transboundary rivers.

As will be seen in the following chapters, this complicated mix of actors in the SPSI framework, together with the fact that our cases included transboundary rivers and our attempts at twinning river basins created a challenging combination of interfaces, presented special problems and demanded special solutions. The first presentation of the case areas is provided below; more details of their characteristics and the problems facing water management in these basins are provided in further chapters of the book.

The Glomma Basin (Norway)

Glomma is the largest river in Norway, located in the eastern parts of the country. The river basin (41,200 square kilometres) is the most populated river basin in Norway, although the population density is low by global comparison. The north-western parts of the Glomma Basin consist of high mountain areas (Jotunheimen area, with Norway's highest peak: Galdhøpiggen, at 2468m above sea level), with high precipitation and glaciated areas. The eastern part is covered by forested areas, whereas the central and southern regions comprise large agricultural areas. The large side branch entering the Glomma River between

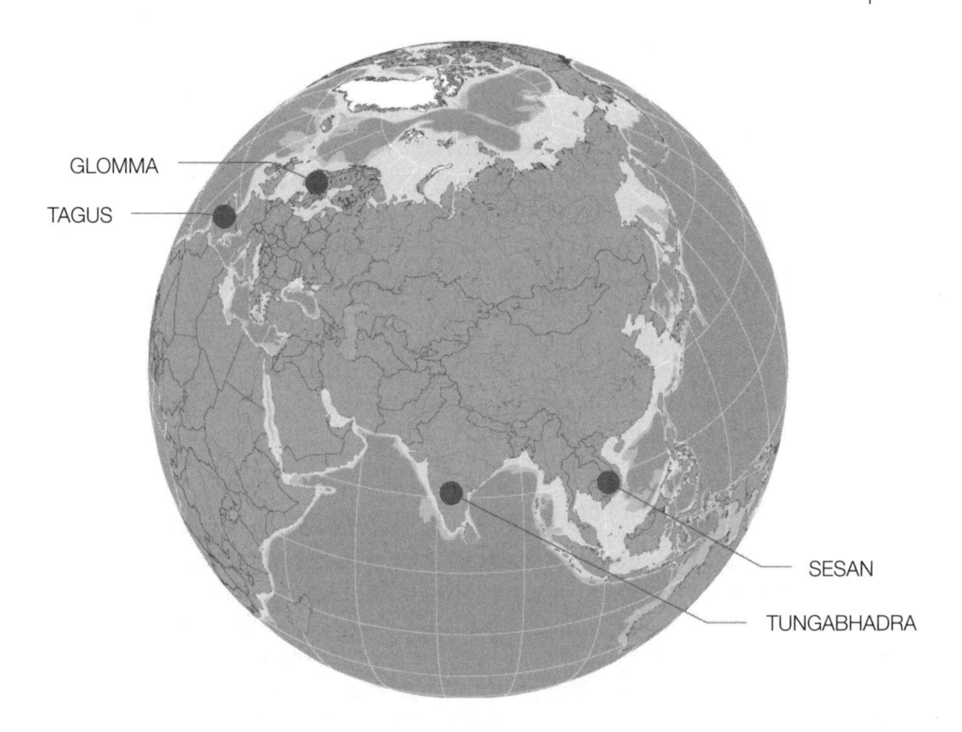

Figure 1.1 The selected four river basins in STRIVER: Tungabhadra (India), Sesan (Vietnam and Cambodia), Glomma (Norway) and Tagus (Spain and Portugal)

Source: L.J. Barkved (www.striver.no)

the lakes of Mjøsa and Øyeren is often referred to as the Laagen Watercourse. The Glomma contains Norway's largest lake, Mjøsa, which has a surface area of 350 square kilometres, and a maximum depth of 450m. At the entrance of Lake Øyeren, the River Glomma forms northern Europe's largest inland delta, the Northern Øyeren Nature Reserve, which is regarded as an extremely important wetland for migratory water fowl (a Ramsar site).

The Glomma and Laagen River Basin features 56 hydropower stations and 26 regulations, and water and diversion schemes. Overall, storage comprises approximately 3500 million cubic metres. Hydropower development has a long historical tradition where ecological considerations have been taken when minimum flow requirements (i.e. environmental flow) have been set by the authorities. The basin has, in addition to hydropower development, clearly been influenced by human impact. The influence is typical for industrialized countries and includes land-use changes (new farmland, deforestation, flood embankments, housing, industrial development, infrastructure, etc.) and pollution. The pollution situation in the river has improved considerably over the last 25 years. However, there are still many small settlements that do not have efficient effluent treatments. Fishing, both commercial and leisure fisheries, are considered as important. The regulation and the pollution discharges have created problems for these activities. Large amounts of money have been invested in abating the eutrophication of Lake Mjøsa and Lake Øyeren. This abatement has been successful; but a new pollution problem has arisen in the form of environmental toxins entering the food chain, making it risky to consume fish. The many regulations have also negatively affected fisheries in numerous places.

In this book, Glomma will be form the basis of twinning case studies on environmental flow methodologies (twinned with Sesan; see Chapter 7) and pollution load modelling (twinned with Tungabhadra; see Chapter 5). The well-established procedures for public participation will also be explored (see Chapter 3) in addition to experiences with IWRM.

The Tagus Basin (Spain and Portugal)

The Tagus River (80,100 square kilometres) is located in the centre of the Iberian Peninsula and flows from east to west for 1009km (73 per cent of its length is in Spain, 27 per cent in Portugal). The Tagus River is one of the main water sources in Spain and Portugal and is used for both urban (i.e. the city of Madrid) and agricultural purposes. The climate in the basin is of Mediterranean type, with strong continental influences. The average annual precipitation varies a great deal in space and time. The location of Madrid, Lisbon and other cities in the Tagus Basin is of great importance to the domestic use of water. In the case of Madrid, there is a complex network of reservoirs and canals that ensure there is a sufficient water supply for more than 5.3 million inhabitants. The volumes of water for industrial waste are very low compared with other basins in northern Spain. The Tagus is heavily affected by the construction of large reservoirs.

There are a total of 40 reservoirs with more than 15hm^3 of capacity in the Tagus Basin, which affect most of the main tributaries. Close to the border between Spain and Portugal is the Alcántara Reservoir (storage capacity of 3162hm^3), which is the largest in Europe and significantly modifies the downstream river regime.

Overall, the Tagus River faces a number of complex strategic problems, given the conflicting concerns between water and landscape conservation and increasing water demands within the scenario of global change. The Tagus Basin has many reservoirs with different purposes (irrigation, urban supply and hydropower production). Irrigation is the purpose of many communities and farms. There is increasing pressure on water resources as more than 6 million inhabitants depend on the discharge of the Tagus River and its tributaries to ensure their water supply. Changes in water resources would directly affect the complex management of reservoirs and would make it necessary to raise again questions regarding the use of water within and outside the basin. This river basin is therefore important as a 'transboundary case' in Europe, where there are a large number of such river basins. It highlights the pure scientific challenge in terms of water use and allocation, water management, pollution, land-use changes and the effects of these changes on a river basin.

The Sesan Basin (Cambodia and Vietnam)

The Sesan River is one of the largest tributaries of the Mekong River and has a drainage area of 17,100 square kilometres: 11,000 square kilometres in Vietnam and 6100 square kilometres in Cambodia. With its origin in the central highlands of Vietnam and the southernmost part of Laos, the river flows through mountainous areas in Vietnam's Dak Lak, Gia Lai and Kon Tum provinces before entering north-east Cambodia, where it moves into relatively lowland areas. In Cambodia, Sesan winds from east to west through Ratanakiri Province and into Stung Treng Province, where it merges with the Srepok River, another large tributary of the Mekong River. The resulting large river flows east, where it flows into the Se Kong River just before this river enters the Mekong River close to Stung Treng town. The rainy season lasts from August to November, with peaking flow normally in September to October. The precipitation varies from approximately 1000mm per year in the lowlands in Cambodia to 2200mm per year in the highlands of Vietnam. There are two major cities on the Vietnamese side of the border: Kon Tum (population of 13,800) and Plei Ku (population of 170,000). On the Vietnamese side, the river basin is situated in Kon Tum and Gia Lai provinces. The population density in these two provinces is 32 and 71 people per square kilometre, respectively (*Statistical Yearbook of Vietnam*, 1999).

After the building of the Yale Hydropower Dam between 1996 and 2000, its regulation regime led to flow change downstream, seriously affecting water use and ecosystem health for both Cambodia and Vietnam. Resulting meetings and conferences consisted of representatives from both Vietnam and Cambodia, who worked on how to solve the problems of conflicting river use. Both countries are also members of the Mekong River Commission, which is a cooperative forum for both the utilization and protection of the Mekong River system and its tributaries. Both Vietnam and Cambodia have to implement modern principles regarding water management to cope with the development of more intensive and large-scale water resources exploitation. In both countries this new water management organization has started; but there is still a long way to go to be able to secure the 'rights' of all water-use interests, as well as a healthy aquatic environment. The Sesan case study is important as a 'transboundary case' in Asia where there are a large number of such river basins.

The Tungabhadra Basin (India)

The Tungabhadra River is located in the southern parts of India. Tungabhadra is the largest tributary of the river Krishna, contributing an annual discharge of 14,700 million cubic metres at its confluence point to the main river. It has a drainage area of 71,417 square kilometres, out of which 57,671 square

Figure 1.2 Fishermen drying fish on the banks of the Tungabhadra

Source: U.S. Nagothu

kilometres lie in Karnataka State, after flowing for a distance of 293km; the remaining catchment is found in the state of Andhra Pradesh. Tungabhadra is the lifeline of Bellary, Koppal and Raichur districts in Karnataka and Kurnool, Ananthapur and Cuddappah in Andhra Pradesh.

The basin is mostly rain fed, dominated by red soils, and the average annual rainfall is 1200mm. The major crops grown are paddy, jowar, sugarcane, cotton and Ragi (millets). Prior to the development of large dams and reservoirs in the downstream regions of Tunga and Bhadra sub-catchments, comprised mostly of arid and semi-arid regions, water management had reached a high level of sophistication, both for surface as well as groundwater utilization in agriculture. Tungabhadra Reservoir has been constantly losing its water storage capacity over the decades due to the accumulation of mud as a result of mining, dust, soil erosion and debris, much to the concern of governments. The amount of rainfall has also decreased in the past few years; since the reservoir is not readily filled up, water is released for only one crop.

Conflicts have arisen between Karnataka and Andrah Pradesh due to increasing storage capacity and water use in the upstream part of the basin; but in lean years, in terms of rainfall and river flow, none or very little water reaches Andrah Pradesh. The river catchment includes a number of industrial activities in small and large plants and a wide range of commercial agricultural activities.

The Tungabhadra case study is important as a 'transprovincial case' in India where there are a large number of similar river basins. It highlights the pure

Table 1.1 Water management issues in the case basins

Issues	**Tagus**	**Glomma**	**Tungabhadra**	**Sesan**
Location	Spain/Portugal	Norway	(Karnataka/ Andhra Prasesh,) India	Cambodia/ Vietnam
Drainage area (km^2) and annual precipitation (mm)	81,000km^2 450–1100mm	41,200km^2 260–1050mm	71,417km^2 450–1200mm	17,000km^2 1000–2200mm
Major water uses in the basin	Hydropower, agriculture, domestic use, irrigation, tourism, industry	Hydropower, fishing, tourism, agriculture, industry	Agriculture, domestic use, hydropower	Hydropower, agriculture, fishing
Major land use (in order of priority)	Agriculture, urban settlements	Forestry, agriculture	Agriculture, urban settlements, forestry	Agriculture, forestry
Environmental data, information availability and sharing	Ok, but not shared properly, especially between the two countries	Good and shared well in relation to the other three basins High public access	Not well monitored and not shared between the two provinces	Poor data collection; poor sharing within and between the two countries
Major problems	Water use and allocation (scarcity), transboundary conflicts, pollution droughts, fires	Pollution-refined environmental flow assessment needed	Water use and allocation, water scarcity, pollution, transboundary conflicts deforestation	Transboundary problems regarding water use and allocation; lack of pollution assessments
Hydropower projects (HPPs)	Major hydropower projects set up on the river Water transfer	Series of 56 hydropower stations and 26 reservoirs	Two major hydropower projects	Major hydropower projects (several planned HPPs)
Policy	Lack of basin-level IWRM policy framework	Well-formulated basin-level policy	Lack of basin-level IWRM policy framework	Lack of basin-level IWRM policy framework

scientific challenge in terms of water use and allocation, water-use conflicts, pollution, land-use changes, and the effects of these changes on the river basin. However, the socio-economic aspects of the case are of very high relevance in a country such as India, where the differences in standards of living between the various social classes are tremendous and the country is in the middle of an accelerating change. The case study research identifies procedures for enhanced end-user involvement and public participation in order to develop sustainable collective action in the water sector at all levels of political organization.

The challenges facing water management in the basins

While these case studies are described in detail in the coming chapters, it may help the reader to understand the overall challenges by introducing them here. As can be seen from Table 1.1, both the size of the basins as well as annual precipitation differ considerably: the Tagus, which flows through much of the Iberian Peninsula, has the largest drainage area (81,000 square kilometres), while the Sesan has the smallest (17,000 square kilometres). Considering major water uses, all of the rivers are used for hydropower as this was one of the criteria in the choice of case basins; we were especially interested in water management in heavily modified rivers. Not surprisingly, water is used for irrigation and agricultural purposes in all four river basins; agriculture accounts for around 80 per cent of water use in many parts of the world as considerable amounts of water are necessary for food production. What cannot be seen in Table 1.1 is the amount of precipitation that evaporates – that is, the degree to which rainfall can be used in one way or another. In all basins competing uses of water are a major problem, and this is especially problematic in three of the basins as these do not have functioning integrated water management systems in place. In all of the basins, except the Glomma in Norway, a lack of data, or of data-sharing, increases the problems of management. As we demonstrate in the coming chapters, these problems with data-sharing and with the collection of different forms of data necessary for sustainable water management are a major challenge in which the SPSI plays a central role. It is our hope that the descriptions of the work carried out in the four case basins included in this book will provide a contribution to improving the SPSI and, therefore, to water management in the basins, as well as in other places in the world.

References

Bates, B. C., Kundzewicz, Z. W., Wu, S. and Palutikof, J. P. (eds) (2008) *Climate Change and Water*, Technical Paper of the Intergovernmental Panel on Climate Change, IPCC Secretariat, Geneva

General Statistics Office of Vietnam, www.gso.vn, last accessed March 2010

Gooch, G. D. (2004a) 'Improving governance through deliberative democracy: Initiating informed public participation in water governance policy processes', Paper presented to the 14th Stockholm Water Symposium, 16–20 August 2004, Stockholm

Gooch, G. D. (2004b) 'The communication of scientific information in institutional contexts: The specific case of transboundary water management in Europe', in S. Laangnäs and J. Timmerman (eds) *Environmental Information in European Transboundary Water Management*, IWA, The Netherlands

Gooch, G. D. and Huitema, D. (2007) 'Participation in water management', in J. G. Timmerman, C. Pahl-Wostl and J. Möltgen (eds) *The Adaptiveness of IWRM: Analysing European IWRM Research*, IWA Publishing, London

Gooch, G. D. and Stålnacke, P. (eds) (2006) *Integrated Transboundary Water Management in Theory and Practice: Experiences from the New EU Eastern Borders*, IWA Publishing, London

Jenkins, G. J., Murphy, J. M., Sexton, D. M. H., Lowe, J. A., Jones, P. and Kilsby, C. G. (2009) *UK Climate Projections: Briefing Report*, Met Office Hadley Centre, Exeter, UK

Klijn, E. H. and Koppenjan J. F. M. (1997) 'Beleidsnetwerken als theoretische benadering: Een tus-senbalans', *Beleidsnetwerken*, vol 2, pp143–167

Kuhn, T. S. (1970) *The Structure of Scientific Revolutions*, University of Chicago Press, Chicago, IL

Mostert, E. (1999) 'Perspectives on river basin management', *Physics and Chemistry of the Earth Part B: Hydrology, Oceans and Atmosphere*, vol 24, pp563–569

Roux, D. J., Rogers, K. H., Biggs, H. C., Ashton, P. J. and Sergeant, A. (2006) 'Bridging the science–management divide: Moving from unidirectional knowledge transfer to knowledge interfacing and sharing', *Ecology and Society*, vol 11, no 1, p4, www.ecologyandsociety.org/vol11/iss1/art4/, last accessed March 2010

Statistical Yearbook of Vietnam, (1999), www.gso.gov.vn/default_en.aspx?tabid=515&idmid=5&ItemID=5691, last accessed March 2010

Turton, A. R., Roux, D., Claassen, M. and Hattingh, J. (eds) (2006) *Governance as a Trialogue: Government–Society–Science in Transition*, Springer-Verlag, Berlin

Vörösmarty, C. J., Green, P., Salisbury, J. and Lammers, R. B. (2000) 'Global water resources: Vulnerability from climate change and population growth', *Science*, vol 289, no 5477, pp284–288

2

Putting the 'Integration' in the Science–Policy–Stakeholder Interface

Bruna Grizzetti, Fayçal Bouraoui,
Geoffrey D. Gooch and Per Stålnacke

Introduction

The analysis of the interactions between science, policy and stakeholders in water management leads us to consider the process of information exchange and knowledge production occurring at their interfaces. This process has often been addressed as integration. In water management, the idea of integration is linked to the flow of information between the involved actors, the reciprocal understanding of needs and perspective, and the production of new knowledge, including different expertise and potential competing interests, within the framework of a shared commitment to enhance water resource protection and management.

In this chapter we analyse the role and the mechanisms of integration in science–policy–stakeholder interactions and we reflect on the type of research promoting such integration to support sustainable water management. We back up this analysis by providing some practical experiences from the STRIVER project (2009), showing some of the difficulties and successful outcomes of integration.

Integration as a key challenge in sustainable water management

The World Commission on Environment and Development (WCED) defines sustainable development as development that 'meets the needs of the present without compromising the ability of future generations to meet their own needs' (WCED, 1987). This concept was adopted in the declaration of the United Nations Conference on Environment and Development (UNCED), held in Rio de Janeiro in 1992, which formally asserted a worldwide commitment to shape global, national and local development strategies according to sustainable development principles (UNCED, 1992). To steer policies' progression towards sustainable development, UNCED adopted a global plan of action, Agenda 21, which promotes the integration of decision-making processes and wider community involvement. Regarding water, Agenda 21 emphasizes the need for an integrated approach to the planning and management of water resource. Agenda 21 includes the 'four Dublin Principles', formulated at the international Conference on Water and Environment in Dublin (1992), which are considered the guiding principles underpinning integrated water resource management (IWRM). These principles stress the need to consider the linkages between environment, society and economic systems for the sustainable management and use of water as a resource, as well as the fundamental role of stakeholder involvement and a participatory approach to water management. Thus, the core of IWRM lies in the 'integration' of different environmental aspects involved in the water cycle and the associated disciplines, and the integration of the environment with social and economic systems.

With the renewed European Strategy for Sustainable Development (EU SDS), the European Union has reinforced its commitment to: shaping its policies and activities to ensure the protection and improvement of the environment; to promote a democratic, socially inclusive and cohesive society; to encourage economic prosperity; and to support sustainable development worldwide (Council of the European Union, 2006). The EU SDS identifies seven key challenges for the implementation of sustainable development policies affecting the environment, economy and society. These include climate change and energy, transport, consumption and production, natural resources, public health, social inclusion, and global poverty and development. In this framework, water protection and management is a highly relevant cross-cutting issue involved in almost all of the key targets.

Concerning the conservation and management of a water resource, the EU SDS affirms that the European Commission and the member states should work towards improving IWRM in all in-land, marine and coastal waters.

But what does integration mean in this context? How can it be realized in the interactions between science, policy and stakeholders? And what type of research is needed to support such integration? The next section will address these questions.

Interdisciplinarity, transdisciplinarity and sustainability science

Government and research financiers have increasingly recognized that multidisciplinarity is required to improve water resource protection and management. Consequently, they have put increasing emphasis on integrated research as a means of addressing the complexity of the problem, involving environmental, economic and social systems as a whole, rather than single disciplines. However, integration does not automatically result just from putting together contributions from different disciplines on the same issue, but rather from an attempt at interdisciplinarity. In fact, interdisciplinarity builds upon the existence of disciplines and develops new framings of knowledge with innovative or combined languages and approaches (Dixon and Sharp, 2007). By addressing real world problems and incorporating greater public accountability, interdisciplinarity responds to the new expectation of scientific research to be more relevant and more responsive (Lowe and Phillipson, 2006).

In addition, some authors have proposed the concept of transdisciplinarity (Pohl, 2005; Nowotny, 2009), to indicate a type of research that takes into account the complexity of the issue, addresses the perceptions of both science and society, produces practically relevant knowledge and includes the interests of the different stakeholders (Pohl, 2005). Transgressing the boundaries between disciplines, the quality control of such research needs to shift from the value added (scientific excellence) within a discipline to value integrated, which includes something of societal value to define good science (Nowotny, 2009). Pohl (2008) defines transdisciplinarity as a process of coproduction of knowledge, where multiple disciplines and stakeholders of other sectors of society are involved in a collaborative process of knowledge production. Four main cultures interact in this process: the bureaucratic, the academic, the economic and the civil policy cultures, which all hold particular perspectives and different visions on the use and role of science and technology. Pohl (2008) identifies two main types of knowledge production in transdisciplinary research. In type one transdisciplinary research, the knowledge produced from different disciplinary fields is reorganized according to the perceived demands of the audience, who are not directly involved in the process. The academic policy culture defines the audience interests and considers itself responsible for reorganizing knowledge and passing on information. In type two transdisciplinary research, the four policy cultures participate in the coproduction of knowledge and the academic policy culture is responsible for organizing the process, but the production of knowledge stays as a 'collective endeavour' of all the policy cultures involved. Pohl (2008) argues that type two transdisciplinary research is the most appropriate to use as a bridge between science and policy.

Recently, the Millennium Ecosystem Assessments (MEAs) have stressed the importance of understanding the interactions between human and natural systems in the context of changing drivers, such as climate change, land-use change, population increase, pollution and economic growth, from the local to the global scale (MEA, 2005). It has highlighted, however, that the existing knowledge gaps cannot be addressed by traditional disciplines, but requires a new kind of interdisciplinary science, which is driven by problem-solving objectives (Carpenter et al, 2009) and embeds in its evaluation an adaptive management approach (Steffen, 2009). This new type of knowledge production is referred to as sustainability science (Kates et al, 2001). Sustainability science seeks to understand the interactions between nature and society, addressing the dynamics of the interactions, the long-term trends, the vulnerability and resilience of the nature–society system, and the opportunities for adaptive management and societal learning (Kates et al, 2001). It promotes both useful knowledge and informed action, resulting in neither basic nor applied research, but rather 'use-inspired basic research' (Clark, 2007).

Figure 2.1 Crossing academic borders: Hydroelectric power is explained to a social scientist by an engineer

Source: K. Völker

The term 'transdisciplinary research' has been used as a synonym for 'sustainability research', as 'transdisciplinarity research addresses the knowledge demands for problem-solving in complex societal concerns with regards to common good' and for development to be sustainable it needs to consider social, ecological and economic factors (Hirsch Hadorn et al, 2006). In fact, transdisciplinarity is a form of interdisciplinarity that emphasizes the need for cooperation between the academic disciplines and the different parts of society, and for this reason it appears particularly adapted to tackle problems related to sustainable development, where the integration between natural science and social science is a key point (Tappeiner et al, 2007).

Integration and the science–policy–stakeholder interface

The discussion on the characteristics of integration and on the type of research needed for implementing sustainable water management helps to clarify the context in which we analyse integration within the science–policy–stakeholder interface (SPSI). This analysis requires a reflection on the role of each actor involved in knowledge production and on the final objective of the process. From this perspective, scientists, policy-makers and stakeholders are all actors involved in the IWRM process. Mostert and Raadgever (2008) suggest that researchers who want to optimize their contribution to water management practice should reflect on the role of their research, analyse the stakeholders involved, choose whom and what to serve, and decide on a strategy, implementation and communication plan.

For the implementation of IWRM, environmental and social research has to come together and integrate with the other parties involved (i.e. society, policy and management), which all have to interact with each other. In addition, in IWRM within environmental research, experts of different fields, such as hydrologists, biologists, chemists, engineers, agronomists, etc., are called on to work together towards a common goal. Hence, there is a clear requirement for interdisciplinarity research. Moreover, as this type of research is problem oriented and involves stakeholders, then this type of research can be considered transdisciplinary as well.

To succeed, the integration of scientific research with policy, management and society has to be considered at all the stages of IWRM development, including design, planning, implementation and review. However, integration needs to be planned at such a high level, even in terms of policy context, funding system, institutions concerned, stakeholders' participation and research fields involved, that it is difficult to attain. More often than not this results in the integration of research fragments in a wider discontinuous scenario. In fact, although required by all recent European water laws, the effective integration of science and policy and of natural scientists and social scientists is not immediately achievable (Pohl, 2005; Quevauviller et al, 2005; Borowski and Hare, 2007; Mee et al, 2008). For example, the Water Framework Directive and the new Marine Strategy Directive promote the integrated and sustainable management of a water resource by taking into account stakeholder views, and require natural and social experts to work together (Hodgson and Smith, 2007). However, while the scientific requirements are clearly defined, those related to stakeholder involvement are quite vague (Fletcher, 2007), as will be discussed in Chapter 3, and the interpretation of standards such as 'good environmental status' enforced by the directives is definitely influenced by human values (Mee et al, 2008). Nevertheless, when addressing issues related to sustainable development, natural and social scientists may find common ground as they both see the ecological relationships as the product of humans and other organisms interacting with the environment. They have started to close the gaps between ideals and practice and to cross the disciplinary boundaries to establish new forms of science that can provide alternative development strategies (Quinlan and Scogings, 2004).

An effective integration at the SPSI passes necessarily and primarily through the identification of common goals and shared motivations. However, one must be aware of many other factors that could prevent or lessen integration. First, the difficulties in communication as different communities use specific terminology and 'jargon' that can constitute a barrier to reciprocal understanding and constructive dialogue. They often also hold different views of exactly what science is (epistemology), and how the world is constituted (ontology). Second, there is the different time frame, which is related to the respective interests, objectives and funding system. In fact, scientific research is more oriented towards a long-term view, while policy and management generally need to act in the short or medium term. Thus, while research in sustainability and IWRM needs to be carried out from a long-term perspective, the horizon of political actions is unfortunately often short term, stretching from one set of elections to the next. Conversely, society is concerned about both the short- and long-term effects; but its priorities may be different from those of scientific research. Finally, in scientific research, recognition, career and funding systems, which influence motivations, may press towards scientific excellence, founded in basic research, and do not always contribute to integrated results and policy relevance.

However, in water management there is a clear need for a 'science–policy interface', where scientific information is streamlined and policy-makers can clarify their questions and needs. Examples in this direction are the Water Framework Directive (WFD) Common Implementation Strategy (CIS), the Pilot River Basin Network and the Harmonica Initiative (Quevauviller et al, 2005). Similarly, researchers and water managers have to overcome their mutual misunderstanding, due to structural and attitudinal differences. They also have to learn to develop effective communication strategies that allow the mutual exchange of information and knowledge. In fact, researchers see integration as a means of studying the complex interactions within the river basins, while water managers are more likely looking towards integrating cost-effective measures within river basin plans (Borowski and Hare, 2007). A great contribution to integration at the science–policy interface may be provided by boundary organizations, such as governmental agencies, which work at the interface and may act as potential productive institutions to promote science–policy interactions (Huitema and Turnhout, 2009). However, in order to avoid conflicts of interest, the role and field of action of such organizations should be clearly framed.

Experiences of inter/transdisciplinarity in the STRIVER project

In IWRM the integration process should be embedded within the management structure, as the level of collaboration and integration are central to the practical implementation. This is, indeed, critical, as even when designed according to interdisciplinary criteria, a management plan may fail to achieve integration since it may promote a general commitment to act collaboratively without developing any new integrated outcomes. In the STRIVER project upon which this book is based, the twinning between researchers from different countries and the integration between different disciplines were constitutive elements of the project objective, planning and management structure. We have tried to analyse the integration achieved during the practical implementation of the project – in particular, that of environmental science with the other disciplines. Some of the

Table 2.1 Type of interdisciplinary engagement

Type of interdisciplinary engagement	Description of interdisciplinary work
Information exchange	Joint meetings are organized to enable an understanding of the questions and of the different perspectives. Separated disciplinary packages are developed in parallel. Joint outputs are collated rather than integrated. The outcomes are mainly discipline-specific reports.
Collaboration	Joint meetings are organized to enable an understanding of the questions and perspectives. Researchers commit to acting collaboratively. The leaders of the different disciplinary teams negotiate tasks and outputs. A few joint outputs are led by different disciplinarily teams with negotiated input from other teams.
Active collaboration	The research is framed by specific disciplinary teams; but the framing is discussed and adjusted iteratively, thus facilitating ongoing exchange and joint paper generation throughout the research process.
Joint research	Research is designed collaboratively. Questions, language, methods, interpretations and implications are fully negotiated. Multiple joint research outputs are generated.

Source: Adapted from Dixon and Sharp (2007)

project partners were asked to express their perception of integration in the project according to four different levels of 'interdisciplinary engagement' (see Table 2.1) by providing practical examples of work and outcomes from the project. The reflection is not systematic and does not necessarily represent the consortium view. However, it is an attempt at self-analysis and review.

In STRIVER, the challenges involved in the practical implementation of interdisciplinarity emerged early on. In fact, according to the project plan, a wide collection of information concerning natural science, socio-economic conditions, institutions, water uses, policy and legislation was planned for the four river basins under study and held in a centralized database. Therefore, a web data repository platform was designed to aid the smooth collection and exchange of information. The collection of the data revealed how natural and social scientists speak differently although using similar words, notwithstanding the sincere and consistent effort to communicate. The experience has shown that only dialogue can improve the reciprocal understanding at this stage, in addition, of course, to a constructive and open attitude. The integration at this stage of the STRIVER project resulted in *information exchange* (see Table 2.1), but provided two important outcomes. First, it provided a concrete and multidisciplinary collection of all the relevant information for the river basins IWRM. Second, it raised awareness of the need to clarify terminology and simplify concepts to facilitate reciprocal understanding. This means that in an integrated project, time and resources need be allocated to this task too. Moreover, one of the main lessons was that defining common objectives may act as a convergence point between the different disciplines. The effort required to explain the respective methodologies, and to share concepts, improved both understanding and actual integration.

Much more experience on integration in the SPSI was then gained through the river basin modelling with stakeholder involvement. According to the STRIVER plan, water-diffuse pollution status was studied in the Glomma (Norway) and in the Tungabhadra (India) river basins through the application of a hydrological model and, in parallel, a stakeholder analysis was conducted (Grizzetti et al, 2008). In principle, hydrological modelling offers a scientific base for water resource assessment and projections of future conditions, while stakeholders' involvement provides the evaluation of current challenges and future options for water uses by considering the competing interests and issues at stake. In order to implement IWRM, the idea was thus to enhance exchanges between scientists and stakeholders. The potential of the interaction and its practical implementation became clearer in the course of project development. In fact, during the modelling process stakeholders could contribute first to prioritizing water quality and quantity problems, targeting the modelling objectives; then to describing more correctly the system, providing input and information based on real experience; and, finally, to building scenarios of real interest and analysing them (see Chapter 5). As a result, the modelling estimates on water current and future pollution were more realistic, understandable and acceptable for the stakeholders involved, and, thus, more relevant to sustainable water management.

However, while simple in principle, during the interaction between scientists and stakeholders in the modelling process and scenarios development, several difficulties were encountered in practice due to differences in priorities, terminology and approaches. An additional practical difficulty was linked to the frequency of the meetings, resources available and the physical distance between modellers and stakeholders (in STRIVER, modellers and stakeholder were, in some cases, from different continents). These aspects illustrate that project flexibility in adapting project resources, time schedule and deliverables is crucial to enhancing the synergies between scientists and stakeholders. As a general perception the level of integration achieved resulted in *collaboration*, but with outcomes going even further, resulting in *active collaboration* (according to the scale presented in Table 2.1) due to the challenge to provide specific inputs or answers from both parts boosting interactions and motivations.

This experience of river basin water pollution modelling with stakeholder involvement can be considered as an attempt towards transdisciplinary research, where scientists and stakeholders have collaborated to coproduce relevant knowledge.

Another opportunity for transdisciplinarity and interaction between the STRIVER researchers and stakeholders was realized by the development of an environmental flow methodology to reduce the negative impacts from hydropower regulation in rivers (see Chapter 7). The aim of the methodology was to set reservoir water release rules to ensure favourable water levels for the river ecology and human uses, within the constraints of economic feasibility. According to this method, first scientific inputs from river ecologists, hydrologists, environmental scientists and economists are used to produce the basic knowledge about the river basin. Then an expert panel is established where the scientists, non-scientific experts and stakeholders jointly:

- define policy-relevant alternatives;
- identify river ecological and user interests (for which impacts are to be determined);
- draw optimal water-level curves for each ecological value and user interest chosen;
- discuss and develop pressure-impact curves for various flow regimes.

Finally, the results are translated into a multi-criteria tool that pools this information and are discussed with stakeholders. The methodology itself requires a high level of collaboration and the outcome of the work definitively integrates the knowledge of different expertise and the trade-off of contrasting interests. In the STRIVER project, the methodology was successfully developed and tested in the Glomma and the Sesan river basins (Berge et al, 2008). After the first stage of sharing, understanding and setting the methodology (a sort of learning or convincing phase that was rather lengthy), the research produced an *active collaboration* with almost *joint research* outcomes (according to the scale presented in Table 2.1). The positive result may be ascribed to the constructive collaboration between the different experts and stakeholders involved and to the integrated nature of the methodology itself. This experience suggests that in IWRM research projects, the choice of opportune methodologies may be beneficial for integration, although time and resources are also needed as a prerequisite for integrated outcomes.

The development of IWRM in transboundary river basins adds further challenges to the integration issues discussed so far, including often contrasting national water-use interests and priorities. In the STRIVER project a policy analysis was conducted in the context of transboundary water management in the Sesan (Vietnam/Cambodia) and the Tagus (Spain/Portugal) river basins. The study included the analysis of actor networks and the development of scenarios with stakeholders' participation (see Chapters 4 and 8). The analysis of actor networks provides further understanding of the forces influencing water management and the stakeholders involved, while the development of scenarios allows ideas about possible future actions to be reviewed. In both transboundary river basins, the scenarios were developed and discussed with the stakeholders from the neighbouring countries, which might include formal actors, such as government departments at national, regional and local levels, as well as NGOs and local people. During the scenarios development, the challenges facing integration were manifold due to different, and often contrasting, interests within each country and between neighbouring countries. To propose relevant scenarios, policy analysts had to understand issues related to fishery, agricultural farming techniques, hydroelectric power dams and drinking water distribution, and discuss them with government representatives, managers, engineers, fishermen, farmers and local people. Although time was necessary to map all the actors involved, and not all them could be completely represented in the stakeholder groups, relevant scenarios were developed and discussed in two river basins, improving awareness of the consequences of certain actions and increasing the level of public participation and social learning (Gooch and Rieu-Clarke, 2008). The scenarios developed in the project can be considered as an example of real integrated outcomes, produced by *active collaboration* between experts and stakeholders (see Table 2.1), as well as an example of transdisciplinarity research. However, within a transboundary context, successful integration sometimes requires being able to overcome some political barriers. For example, in the case of the Sesan River Basin, a major obstacle was to get the stakeholders from the different countries and provinces to sit down together around a table.

Lessons learned

These examples illustrate how during the STRIVER project the science–policy–stakeholder actors interacted, emphasizing the transdisciplinarity characteristics of the research conducted to support IWRM. Some main lessons can be drawn from these experiences as follows:

- Defining and sharing common objectives motivates the participants' integration.
- Clarifying terminology and methodologies promotes communication and reciprocal understanding.
- Flexibility in and reallocation of initially planned project resources, time schedule and deliverables during the project course increases the opportunities for collaboration and more targeted quality of interdisciplinary outcomes.
- Integration requires time and resources that need to be foreseen. The project plan should allow for this specific task.
- The choice of methodologies requiring different forms of expertise should support the interdisciplinary process.

In addition, some general reflections can be proposed on the meaning and added value of 'working together' at the SPSI on research projects supporting IWRM:

- Work together; do not just divide the work.
- Promote shared working tools, such as indicators and geographic information system (GIS) maps.
- Envisage suitable and integrated project deliverables and outcomes.

- Do not define problems and solutions before the analysis; rather, work together to define and analyse problems and propose local appropriate solutions.
- Twinning provides a platform for working together.

References

Berge, D., Barton, D. N., Dang, K. N. and Nesheim, I. (2008) *Environmental Flows (EF) in IWRM – With Reference to the Hydropower Regulated Glomma River in Norway and Sesan River in Vietnam/Cambodia*, STRIVER Policy Brief No 9, http://kvina.niva.no/striver/Portals/0/documents/STRIVER_PB9.pdf, last accessed March 2010

Borowski, I. and Hare, M. (2007) 'Exploring the gap between water managers and researchers: Difficulties of model-based tools to support practical water management', *Water Resource Management*, vol 21, no 7, pp1049–1074

Carpenter, S. R., Mooney, H. A., Agard, J., Capistrano, D., De Fries, R. S., Díaz, S., Dietz, T., Duraiappah, A. K., Oteng-Yeboah, A., Miguel Pereira, H., Perrings, C., Reid, W. V., Sarukhan, J., Scholes, R. J. and Whyte, A. (2009) 'Science for managing ecosystem services: Beyond the Millennium Ecosystem Assessment', *Proceedings of the National Academy of Sciences*, vol 106, no 5, pp1305–1312

Clark, W. C. (2007) 'Sustainability science: A room of its own', *Proceedings of the National Academy of Sciences*, vol 104, pp1737–1738

Council of the European Union (2006) *Review of the EU Sustainable Development Strategy (EU SDS) – Renewed Strategy*, Annex 10117/06, Renewed EU Sustainable Development Strategy, p29 http:// register.consilium.europa.eu/pdf/en/06/st10/st10117.en06.pdf, last accessed March 2010

Dixon, J. and Sharp, L. (2007) 'Collaborative research in sustainable water management: Issues of interdisciplinarity', *Interdisciplinarity Science Review*, vol 32, no 3, pp221–232

Fletcher, S. (2007) 'Converting science to policy through stakeholder involvement: An analysis of the European Marine Strategy Directive', *Marine Pollution Bulletin*, vol 54, pp1881–1886

Gooch, G. D. and Rieu-Clarke, A. (2008) *Water Regimes in Transboundary Heavily Modified Rivers*, STRIVER Policy Brief No 2, http://kvina.niva.no/striver/Portals/0/documents/STRIVER_PB2.pdf, last accessed March 2010

Grizzetti, B., Lo Porto, A., Barkved, L. J., Bouraoui, F., Deelstra, J. and Joy, K. J. (2008) *Modelling Water Pollution with Stakeholders' Involvement – The Twinned Experience of Glomma (Norway) and Tungabhadra (India) River Basins*, STRIVER Policy Brief No 10, http://kvina.niva.no/striver/Portals/0/documents/STRIVER_PB10.pdf, last accessed March 2010

Hirsch Hadorn, G., Bradley, D., Pohl, C., Rist, S. and Wiesmann, U. (2006) 'Implication of transdisciplinarity for sustainability research', *Ecological Economics*, vol 60, pp119–128

Hodgson, S. M. and Smith, J. W. N. (2007) 'Building a research agenda on water policy: An exploration of the Water Framework Directive as an interdisciplinary problem', *Interdisciplinary Science Review*, vol 32, no 3, pp187–202

Huitema, D. and Turnhout, E. (2009) 'Working at the science-policy interface: A discursive analysis of a boundary work at The Netherlands Environmental Assessment Agency', *Environmental Politics*, vol 18, Issue 4, July 2004, pp576–594

Kates, R. W., Clark, W. C., Corell, R., Hall, J. M., Jaeger, C. C., Lowe, I., McCarthy, J. J., Schellnhuber, H. J., Bolin, B., Dickson, N. M., Faucheux, S., Gallopin, G. C., Grübler, A., Huntley, B., Jäger, J., Jodha, N. S., Kasperson, R. E., Mabogunje, A., Matson, P., Mooney, H., Moore III, B., O'Riordan, T. and Svedin, U. (2001) 'Sustainability science', *Science*, vol 292, no 5517, pp641–642

Lowe, P. and Phillipson, J. (2006) 'Reflexive interdisciplinarity research: The making of a research programme on the rural economy and land use', *Journal of Agricultural Economics*, vol 57, no 2, pp165–184

MEA (Millennium Ecosystem Assessment) (2005) *Ecosystem and Human Well-Being: Synthesis*, Island Press, Washington, DC, p137

Mee, L. D., Jefferson, R. L., Laffoley, D. d'A. and Elliott, M. (2008) 'How good is good? Human values and Europe's proposed Marine Strategy Directive', *Marine Pollution Bulletin*, vol 56, pp187–204

Mostert, E. and Raadgever, G. T. (2008) 'The seven rules for hydrologists and other researchers wanting to contribute to water management practice', *Hydrology and Earth System Sciences Discussions*, vol 5, pp843–864

Nowotny, H. (2009) *The Potential of Transdisciplinarity*, www.interdisciplines.org/interdisciplinarity/papers/5, last accessed March 2010

Pohl, C. (2005) 'Transdisciplinary collaboration in environmental research', *Futures*, vol 37, pp1159–1178

Pohl, C. (2008) 'From science to policy through transdisciplinarity research', *Environmental Science and Policy*, vol 11, pp46–53

Quevauviller, P., Balabanis, P., Fragakis, C., Weydert, M., Oliver, M., Kaschl, A., Arnold, G., Kroll, A., Galbiati, L., Zaldivar, J. M. and Bidoglio, G. (2005) 'Science–policy integration needs in support of the implementation of the EU Water Framework Directive', *Environmental Science and Policy*, vol 8, pp203–211

Quinlan, T. and Scogings, P. (2004) 'Why bio-physical and social scientists can speak the same language when addressing sustainable development', *Environmental Science and Policy*, vol 7, pp537–546

Steffen, W. (2009) 'Interdisciplinary research for managing ecosystem services', *Proceedings of the National Academy of Sciences*, vol 106, no 5, pp1301–1302

STRIVER (2009) *Strategy and Methodology for Improved IWRM: An Integrated Interdisciplinary Assessment in Four Twinning River Basins*, Founded by the European Commission under the Sixth Framework Programme (SUSTDEV-2005-3.II.3.6: Twinning European/Third countries river basins), www.striver.no, last accessed March 2010

Tappeiner, G., Taippener U. and Walde, J. (2007) 'Integrating disciplinarity research into an interdisciplinarity framework: A case study in sustainability research', *Environmental Model Assessment*, vol 12, pp253–256

UNCED (United Nations Conference on Environment and Development) (1992) *Report of the United Nations Conference on Environment and Development*, Rio de Janeiro, 3–14 June 1992, Rio Declaration on Environment and Development

WECD (World Commission on Environment and Development) (1987) *Report of the World Commission on Environment and Development: Our Common Future*, General Assembly Resolution 42/187, 11 December 1987

3

The Science–Policy–Stakeholder Interface and Stakeholder Participation

Alistair Rieu-Clarke, Susan Baggett, Dale Campbell, K. J. Joy and Suhas Paranjape

Introduction

The first two chapters of this book maintained that while contemporary challenges relating to the management of water resources necessitates collaboration amongst a range of groups in society, difficulties remain in securing such collaboration. A greater understanding is therefore needed as to how integration between science, policy and stakeholders can be improved. In this chapter we examine the interface between science, policy and stakeholders predominantly from the perspective of stakeholders. The chapter asks what formal and informal platforms are in place, and what tools and methods have been used in order to promote stakeholder participation in water resources management. In addition, the chapter critically reflects on experiences of stakeholder participation within the four case study areas mentioned in Chapter 1.

While the previous chapters have explained why there is a need to focus more attention on understanding the science–policy–stakeholder interface (SPSI), an issue that warrants further consideration here is why stakeholders should be included. A number of reasons justifying stakeholder participation can be identified. At its basic level, stakeholder participation helps to raise awareness about issues that may affect the public. In addition, participation may provide more information for decision-making because citizens and communities know the environment in which they live. Taking account of these additional perspectives and considerations can therefore provide decision-makers with additional options or more creative solutions (Tilleman, 1995, p337; Bruch and Filbey, 2002, p5). Effective participation is also likely to lead to a greater acceptance of the decisions that are taken (Mostert, 2003, p180). While greater acceptance can partly be attributed to better informed and more creative decision-making, a perception that the decision is the result of a legitimate process will increase the likelihood of adherence by those affected (Cash et al, 2002, p2). A further benefit of participation is *social learning*, which relates to 'the growing capacity of social entities to perform common tasks related with a river basin' (Pahl-Wostl, 2008, p484). Promoting social learning can assist 'publics, government and experts' to 'learn how to manage collectively a complex natural resource such as a river basin and deal with conflicting views and interests' (Mostert, 2003, p181).

The necessity of stakeholder participation, to a great extent, is also related to the very nature of water as an ecosystem resource and how it interfaces with the socio-political and institutional context within which it is appropriated. More specifically:

- Water is divisible and amenable to sharing.
- Water is a common pool resource.
- Water has multiple and competing uses, users and resultant trade-offs.
- Exclusion costs are very high.
- Water has multiple scales and boundaries.
- The way in which water is planned, used and managed causes externalities – both positive and negative, and often unidirectional.

The first three characteristics and the characteristics of multiple and nested scales and externalities provide the ground for the various types of stakeholders and highlights their complex and multilevel interaction. The problem of excludability means that simple state-centred solutions, which often involve possible exclusion as an implicit penal measure, will not always work very well in the case of water and highlights that what is needed is much more a process of formation and acceptance of an agreement between different stakeholders. All of these together point to stakeholder participation as an important component and requirement of IWRM.

While recognition of the benefits of stakeholder participation has led to a greater awareness of the need to incorporate a range of perspectives within water management, 'there is still a considerable lack of scientific background, methodologies and tools to support these tasks' (Antunes et al, 2009, p932). Others have noted that there are 'no clear-cut procedures for PP [public participation] that can be followed under every specific context and level' (Özerol and Newig, 2008, p653), and a lack of clarity as to the 'practical meaning of PP' (Mostert, 2003, p179). Part of the challenge is that there is no *one size fits all* solution, and stakeholder participation needs to be flexible enough to adapt to changing factors and circumstances (EC, 2003b). Sometimes the objective of stakeholder participation may also be unclear. Is it, for instance, a means of 'empowering people and enhancing democracy', or 'a marketing tool for government policy' (Mostert, 2003, pp179–180)? Hartley and Wood (2005, p320) observe that 'debate continues about exactly how to undertake public participation and confusion remains about when it should commence, the methods that should be used and which members of the public should be consulted'. Existing knowledge therefore needs to be set within specific contexts in order to bring general discussions of participation to an operational level.

Participation: Defining terms

Prior to considering the formal and informal platforms, and tools and methods for facilitating stakeholder participation, it is first necessary to examine what is meant by stakeholder participation.

What is meant by participation?

In the context of decision-making, *participation* relates to a range of activities that can be broadly classified as information supply, consultation or active involvement. The lowest level of participation involves the provision of information to natural or legal persons (i.e. *information supply*). *Consultation* can be seen as one step further, where relevant bodies actually consult with natural or legal persons to garner knowledge, perceptions, experiences and ideas, although such bodies are under no obligation to account for such information during their decision-making (EC, 2003b, p13). *Active involvement* provides the highest degree of participation, where natural or legal persons contribute both to the development and implementation of decisions and become partly responsible for the outcome (EC, 2003b, p13).

Who participates?

A number of categories are used to classify those who might participate. *Public* participation refers to participation by 'one or more natural or legal persons, and, in accordance with national legislation or practice, their associations, organization or groups' (UNECE, 1998, Art. 2(4)). Renn (2008, p273) provides three categories of the public:

1 The *general public* encompasses 'all individuals who are not directly affected'.
2 The *observing public* is defined as 'the media, cultural elites and opinion leaders who may or may not comment'.
3 The *directly affected public* includes 'individuals and non-organized groups who will experience positive or negative impacts from the outcome of the event or the activity'.

Renn (2008, p273) makes a distinction between the *directly affected public* and *stakeholders*, the latter being 'socially organized groups who are or will be either affected by or have a strong interest in the outcome of the event of the activity'.

Other definitions tend not to make the distinction between socially organized groups and individuals and non-organized groups who are directly affected. For

Figure 3.1 Stakeholder meeting in Pleiku, Vietnam

Source: G. D. Gooch

instance, Article 2(4) of 1998 Aarhus Convention simply refers to 'the public concerned' as being 'the public affected or likely to be affected by, or having an interest in, the environmental decision-making' (UNECE, 1998). Similarly, *stakeholder* or *interested party* has been defined as 'any person, group or organization with an interest or "stake" in an issue, either because they will be directly affected or because they may have some influence on its outcome' (EC, 2003b, p11). Public concerned, stakeholder and interested party may therefore be used synonymously.

By including the term 'having an interest in', the definition of stakeholder could encompass a range of actors, including those directly affected by a decision (i.e. the users, such as farmers, ordinary citizens, industry, etc.), those representing the directly affected (user or trade associations) and those with a general interest (environmental advocacy groups, scientists). As already acknowledged in Chapter 2, when implementing an IWRM project different scientific disciplines do not only interact and work alongside each other to try and identify workable solutions, but also work with the other parties involved. The challenge in contemporary water management is extensive – it involves a wide range of people from a wide range of backgrounds who may be included and defined as a 'stakeholder'. The main challenge in this context is how to fully, and successfully, engage and involve stakeholders who may not be specialists in the 'traditional' disciplines already associated with water management and bring with them into this 'forum' a diversity of knowledge, interests and concerns that need to be taken into account (van Eeten et al, 2003; Loucks, 2006). Moreover, different stakeholders are characterized as having dissimilar perspectives on issues; the central part of problem-solving in this situation is often the differing interests that these actors have in a particular resource or within a natural context (Bouwen and Taillieu, 2004). When stakeholders engage in a 'multi-organizational partnership', they do so for two different reasons – to gain from a collective sense of purpose amongst the stakeholders involved, or as an attempt to resolve differences; in reality, these two reasons often become entangled (Gray, 1989, 2004). Stakeholders may need to reframe their perspective on the purpose of the partnership to take into account other stakeholders' viewpoints (Gray, 2004).

An imbalance of power between different stakeholders, coupled with a logic driven principally by the lead agencies, often leads to problems regarding the inclusion of the weaker actors' emergent voices, resulting in a process far removed from what was initially intended (Bouwen and Taillieu, 2004). A case in point is when the interests of a stakeholder are not properly included during the early stages of a planning process, arising in severe conflicts later on (Janssen et al, 2006). Water disputes between stakeholders or states can arise based on, for instance, economic or social development issues, the control of the water source and political interests; in more severe cases they can culminate in water being used as a military target or tool (Phillips et al, 2006). Disagreement is not always detrimental to a stakeholder process: in some instances it can be of benefit by helping to identify interests, problems, relationships or to initiate change (Priscolli, 2003). It is, however, important to actively manage boundaries in a multi-stakeholder process, as the inception of the process *in itself* creates at least one new boundary (Mostert et al, 2007); it is also crucial that the structure of a participatory process addresses and accommodates both its substantive and procedural aspects in order to provide for different stakeholders (Sharp, 2002; Videira et al, 2006).

In addition, for any given issue, 'stakeholders' may change over time. Individuals may shift responsibilities (e.g. from farmer to policy-maker), or interests may change over time (e.g. from poacher to gamekeeper). Moreover, attitudes and opinions may evolve as a direct result of the participatory process. Identifying stakeholders and managing the participation process is therefore complex. The effectiveness of the methods and tools used for these processes will therefore be considered in more depth later in the chapter.

Legal platform for facilitating stakeholder participation in the SPSI

Stakeholder participation in water resources management is largely contingent on formal rules and principles being adopted and implemented at the national level. Such rules and principles rest on three pillars, as articulated in the Convention on Access to Information, Public Participation in Decision-Making and Access to Justice in Environmental Matters (Brady, 1998; UNECE, 1998; Stec and Casey-Lefkowitz, 2000). While regional in scope, former United Nations Secretary-General, Kofi Annan noted that 'the significance of the Aarhus Convention is global ... it is the most ambitious venture in the area of "environmental diplomacy" so far undertaken under the auspices of the United Nations', which has the potential to 'serve as a global framework for strengthening citizens' environmental rights' (Stec and Casey-Lefkowitz, 2000, p5). The so-called Aarhus Convention was adopted at the Fourth Ministerial Environment for Europe Conference in Aarhus, Demark, on 25 June 1998, and subsequently entered into force on 30 October 2001. As of 11 April 2009, there were 42 parties to the convention across Europe and Central Asia.

In relation to the first pillar, access to information, the Aarhus Convention stipulates that upon a request for environmental information, each party must ensure that public authorities make information available to the public within a reasonable time (UNECE, 1998, Art. 4). The convention provides a limited set of conditions whereby a request may be refused, such as on grounds of public security, interference with the course of justice, legally protected commercial confidentiality, and intellectual property rights (UNECE, 1998, Art. 4). Grounds for refusal must be interpreted in a restrictive manner and communicated and justified in writing (UNECE, 1998). In addition, the Aarhus Convention places obligations on public authorities to possess, update and disseminate environmental information relevant to their functions (UNECE, 1998, Art. 5).

Pursuant to the second pillar, participation in decision-making, the Aarhus Convention stipulates that certain arrangements must be made by public authorities to ensure that the affected public and environmental non-governmental organizations (ENGOs) are able to comment on planned projects, plans, programmes and laws likely to affect the environment (UNECE, 1998, Art. 6–8). Such arrangements include providing information on planned environmental decision-making procedures in a timely and effective manner, promoting participation at the appropriate stages, and obliging authorities to take due account of the outcomes of any public participation.

Legal aspects of participation are closely aligned with the move towards establishing more concrete procedures relating to environmental impact assessments (EIAs) throughout the world. As Howarth (2007, p152) notes: 'EIA is perhaps the first recognition of a public right to be fully informed about the implications of a development project and to have access to appropriate information to enable participation in those procedures that determine the environmental quality of the areas in which ordinary citizens must live.' Through European Commission (EC) and United Nations Economic Commission for

Europe (UNECE) legislation, environmental impact assessments have extended from their initial focus on infrastructure projects to programmes, policies and legislation (UNECE, 2003).

The third pillar of the Aarhus Convention requires each party – within the framework of its national legislation – to provide any person who considers that their right to either information or participation has been violated with a right of redress through a review procedure before a court of law, or another independent and impartial body established by law (UNECE, 1998, Art. 9).

While international law, such as the Aarhus Convention, EC directives and the resultant national legislation, provides a basis by which to facilitate stakeholder participation, at least in Europe, key challenges remain in their implementation. More research is needed to understand precisely how such legal and institutional frameworks affect participation and can be improved. One area in need of further development is the definition of terms and interpretation of provisions. Hartley and Wood (2005), for example, observe that 'the specific meaning of terms contained in the Aarhus Convention, notably "early" and "effective" participation, remain undefined'. Rault and Jeffrey (2008) maintain that 'the single most important problem concerning the implementation of the WFD [Water Framework Directive] is that neither the IWRM nor PP [public participation] has consensual meanings in terms of objectives and priorities'. While vague terminology can enhance flexibility in applying such terms to particular facts and circumstances, a key question remains: whether more specificity would ultimately enhance the implementation of such provisions.

Implementation is also largely related to legal commitments, and is contingent upon the cultural and political context in which participation operates, as well as upon choosing the appropriate tools and methods; these aspects will therefore be considered in more detail in the next two sections.

A related issue concerns the role of law in facilitating stakeholder participation. The Water Framework Directive, for instance, provides an obligation to *encourage* involvement under Article 14 of the WFD. Should such an obligation be more stringent and actually *require* states to involve stakeholders? If so, how can such implementation be monitored and evaluated? This takes us to the implementation of stakeholder participation, and to international and public policy.

International and public policy, participation and the SPSI

Participation has gained increasing recognition within international policy related to water. Within the context of integrated water resource management (IWRM), the 1992 Dublin Statement called for water development and management to be 'based on a participatory approach, involving users, planners and policy-makers at all levels' (Dublin, 1992). Similarly, Agenda 21 advocated for IWRM to be based, in part, on 'an approach of full public participation, including that of women, youth, indigenous people and local communities in water management policy-making and decision-making' (Agenda 21, 1992).

In 2002, the Johannesburg Action Plan reiterated such a call by encouraging the facilitation of 'access to public information and participation, including by women, at all levels in support of policy and decision-making related to water resources management and project implementation' (Johannesburg, 2002).

A similar trend towards emphasizing the need for participation can be seen within international policy related to governance. Indeed, by shifting the emphasis from *government* to *governance*, international policy has sought to place stakeholder participation at the very heart of decision-making processes related to water. The United Nations Development Programme (UNDP) defines *governance* as 'the way a society organizes itself to make and implement decisions – achieving mutual understanding, agreement and action' (UNDP, 2004). Moreover, 'governance' is said to include 'the mechanisms and processes for citizens and groups to articulate their interests, mediate their differences and exercise their legal rights and obligations' (UNDP, 2004). Similarly, the Global Water Partnership (GWP) has sought to define *water governance* as being the range of political, social, economic and administrative systems that are in place to develop and manage water resources, and the delivery of water services at different levels of society (Rogers and Hall, 2003, p7).

Strictly speaking, the GWP definition of water governance does little to suggest the degree to which stakeholders should participate in the decision-making process. However, further insights can be gleaned from international policy related to *good* governance. The European Commission, for instance, notes that 'As the concept of human rights, democratization and democracy, rule of law, civil society, decentralized power sharing, and the sound public administration, gain importance and relevance as a society develops into a more sophisticated political system, governance evolves into good governance' (EC, 2003a). Within the context of water, the 2000 Hague Declaration highlights a key challenge as being 'to ensure good governance, so that the involvement of the public and the interests of all stakeholders are included in the management of water resources' (Hague, 2000).

In summary, this brief review of international policy shows that the need to involve stakeholders within policy processes surrounding water is widely accepted. However, while such acceptance encourages the modalities of *who* participates in *what* activities, the *how* remains largely undefined, at least at the international policy level. The purpose of the next section will therefore be to consider whether more details can be found within existing legal and institutional frameworks.

The political and cultural context for stakeholder participation

There has been a discernable political shift towards recognition of the need for greater stakeholder participation in policy-making (Heere, 2004). Such a shift is reflected in numerous policy statements as outlined in the previous section. A related trend has been the growing emphasis on the need for *good* governance. The crisis over water has even been described as a crisis of governance (UNWDR,

Figure 3.2 Project scientists and local fishermen and women in the Tungabhadra Basin, India

Source: U. S. Nagothu

2006). By shifting the emphasis towards governance, international policy has further reflected the need to involve stakeholders within the management of water resources. Such an emphasis can be seen from the very definition of governance as being 'the way [in which] a society organizes itself to make and implement decisions – achieving mutual understanding, agreement and action' (UNDP, 2004). Moreover, 'governance' is said to include 'the mechanisms and processes for citizens and groups to articulate their interests, mediate their differences and exercise their legal rights and obligations' (UNDP, 2004). As indicated earlier in the chapter, stakeholders bring their own identities (including knowledge and worldviews) to the table, and racial, ethnic, caste and gender aspects may be as important as size of landholding. Thus, stakeholder interaction needs to be seen as a process that is as much structured by the racial, ethnic, caste and cultural-regional identities as much as by their economic stakes. If the Habermasian concept of deliberative democracy that underpins stakeholder participation is to be effective, there must therefore be a 'level playing field' for all the stakeholders.

This is all the more important because of the unequal power relation that exists between and within stakeholders, emanating from unequal access to resources, differential access/closeness to state power, and unequal access to data and information. There may also be deep-seated suspicions amongst some stakeholders because of an historical legacy. For example, in the American context there is much suspicion about the conservation agenda of the conservationists because of the historical legacy and the way in which it affected some of the

Native American tribes/communities because of overbearing conservationists. Thus, it is important to recognize contexts and differentiations and take corrective steps. For example, without positive discrimination and special supportive measures for those who lack political-economic power and voice, multiple stakeholder platforms (MSPs) have been likened to setting up common platforms which only benefit the powerful players. It is here that the principle of equity needs to be strengthened in order to become part of a common reference framework respected by all, with even some kind of qualified veto powers in favour of the disadvantaged. Creating a level playing field also means facilitating the disadvantaged to find their voice, making it heard and developing the staying power needed to then negotiate with others on an equal footing. Such an approach leads back to the need to establish the three legal pillars of stakeholder participation, as described above.

What methods and tools can be used to facilitate interaction within the SPSI context?

Significant efforts have been made to develop and research appropriate tools for different methods of engagement. Welp et al (2006) suggest that the participatory tools available may be divided into two kinds – communication tools that support dialogue and analytical tools which aid assessment and help formalize participants' mental models – and that the way ahead may be to integrate the two. The literature on participation is characterized by a vast array of methods and techniques ranging from 'those that canvass opinions to those that elicit judgements and decisions from which actual policy can be derived' (Gooch and Huitema, 2008, p33). Increasing levels of involvement are still commonly portrayed as a progression, from information provision to collaborative decision-making. However, 'this does not imply that one is more important than the other. The challenge is to select an appropriate level of participation for a particular task' (du Toit and Pollard, 2008, p709).

There has also been an increase in research that has either partly or fully sought to further knowledge and understanding as to how effective participation can be implemented within different river basin contexts. The EC-funded MANTRA-East project (Gooch, 2004), which ran from 2001 to 2004, for instance, had the key objectives of developing methods to communicate and utilize scientific information, and to develop institutional arrangements for public and stakeholder participation. Similarly, River Dialogue (Huitema, 2005) sought to identify the best approaches to increasing public empowerment and the involvement of the public in the implementation of the EC Water Framework Directive. Social learning was the central element of other EC projects – for instance, Harmonizing COllaborative Planning (HarmoniCOP) (Ridder et al, 2005) and Social Learning for the Integrated Management and Sustainable Use of Water at Catchment Scale (SLIM Project, 2004).

From their overview of participatory methods relevant to river basin management (see Table 3.1), Videira et al (2006, p23) maintain that it is of

Table 3.1 Overview of participatory methods

Non-deliberative	Deliberative	
	Random selection of participants	**Stakeholder identification and selection**
Surveys	Focus groups	Advisory committees
Polls	Citizens' juries	Visioning workshops
Public comments	Consensus conferences	Participatory modelling
Public information	Deliberative monetary valuation	Social multi-criteria evaluation
Public hearings	Deliberative polling	Mediation and negotiation

Source: From Videira et al (2006), adapted from Beierle (1998) and Kallis et al (2004)

value to investigate the developing deliberative approaches that encourage open and constructive discussion between interested parties, as opposed to the conventional methods 'mainly characterized by one-way information flows between the public and decision makers'. Abels (2007) also puts forward a typology, linking form (i.e. who participates and how) to function (i.e. why), based on the number and diversity of stakeholder groups involved within the participatory procedure used, resulting in the following basic classification:

- dialogue procedure;
- participatory technology assessment (pTA) in a narrow sense;
- legal public hearing;
- consensus conference;
- extended consensus conference;
- voting conference;
- scenario workshop.

The authority assigned to a participatory arrangement is an important issue. However, the level of authority or power ascribed to a participatory process as a standalone is not enough. There are, as Gooch and Huitema (2008, p32) point out:

> ... other design parameters that are equally relevant. Imagine a participatory process with full-fledged authority to decide on a range of water management issues, but with no information whatsoever? Even though the participants are formally very powerful, lacking any information, they would have to refrain from taking decisions.

Furthermore, 'participatory procedures do not *per se* improve the democratic legitimacy and accountability of policy-making. In order to do so, their linkage to the political system has to be reconsidered and improved – empirically as well as conceptually' (Abels, 2007).

With these considerations in mind – when and what type of participation is appropriate when considering the SPSI in the context of water resource management? When planning participatory processes, it is suggested that the following are taken into account (Videira et al, 2006; Gooch and Huitema, 2008):

- Who should participate?
- What is the timing for participation?
- Which participatory method should be used?
- Which competencies should the participants have?
- What sort of information will the participants have access to?
- What are the financial costs involved?

It is recognized that who and who is not involved, or viewed as having a 'stake' in a participatory process, is both crucial and complex; the process of selecting representative 'groups of stakeholders in river basin management is a highly political one' (Wester et al, 2003, p809). A number of methods only provide for a small group of participants (e.g. citizens' juries, focus groups or social learning groups), while other methods (e.g. referendum, opinion survey, web panel) allow for the inclusion of a greater numbers of participants (Gooch and Huitema, 2008). Selecting a particular participatory method depends on the kind of representation and participant selection procedure required. And 'while some mechanisms rely on the random selection of participants (usually based upon predefined representative criteria), others follow stakeholder analyses to inform the participant identification and selection process' (Videira et al, 2006, p23). The identification and salience of a stakeholder are dependent on the presence, number and combination of three dynamic attributes – power, legitimacy and urgency – which may vary over time or in relation to issues, thereby making a 'critical difference in managers' ability to meet legitimate claims and protect legitimate interests' (Mitchell et al, 1997, p882).

Uncertainty can also act as a barrier to an effective stakeholder process. There are two different types of uncertainty that can affect decision-making:

1 Normative uncertainty – that is, when an actor or group of actors are unconvinced or undecided about their goal and which action they should take in order to fulfil their goal(s).
2 Informational uncertainty, which relates to the informational deficiency of a decision-maker (Newig et al, 2005).

Uncertain aims can sometimes explain the variance between stakeholders' interests and expectations as most stakeholders may deal with uncertainty in decision-making by developing their own mental models and scenarios in order to be able to carry out their task (Vega-Leinert and Schröter, 2008). Reasons for widening participation can therefore be a means of addressing those uncertainties by attempting to reconcile interests and goals; profit from local knowledge; gain insight into a social system; and gain information about the possible acceptance of options, which could lead to better informed and more easily implemented decisions (Newig et al, 2005).

Communication is also a fundamental constituent of participation. Newig et al (2008, p382) argue that an important feature of communication methods in participatory processes and their appropriate selection is the level to which methods are 'formalized', which they define as:

> ... what is relevant in a given context and what is not, and thus constitutes meaning. A range of methods, instruments, tools, and models is available to structure information flows and foster (collective) learning in these participatory processes. Depending on the respective goals and context, more or less formalized methods can be employed. Formalization refers to the extent to which information is channelled in a certain way, leaving more or less scope for open communication.

For example, methods such as the citizens' jury or social learning groups clearly focus on aiding 'multiple flows of information between the public/the stakeholders and experts', whereas a referendum or opinion survey do not (Gooch and Huitema, 2008).

Managing the boundaries between different forms of knowledge and disciplines 'across scales of geography and jurisdiction' is also often critical to transferring information as they provide 'important functions (e.g. protecting science from the biasing influence of politics, or helping organize and allocate authority), but they can also act as barriers to communication, collaboration, and integrated assessment and action' (Cash et al, 2002, p1). Consequently, Cash et al (2002) maintain that the management of these boundaries seems to be fundamental for linking knowledge to appropriate action – and that science and technology interfaces who consciously manage them while also concentrating on balancing the trade-offs between salience, credibility and legitimacy are usually more successful than those who pay no heed.

Scientists and policy-makers in many instances are not detached entities outside the participatory process but are also stakeholders. Steyaert and Jiggins (2007, p584) remark that 'multi-stakeholder learning processes, if adequately conducted, [are] open space for people – including scientists and policy-makers – to speak about their assumptions, values and norms so that decisions become based less on the defence of autonomous interests and hidden meaning and more on appreciation of the interdependency of collective interests'. The extent to which the stakeholders' different knowledge, experiences and actions can be brought together in a management system, however, depends on the capacity of the different actors to accommodate and make sense of each others' domain (Burgess et al, 2000). From their assessment of different participatory models, Abels and Bora (2004) consider two models (voting conference and scenario workshop), where all participating groups enjoy equal procedural rights, as 'balanced models' since policy-makers are integrated on equal terms with other actors, both in the actual deliberative process and the production of knowledge.

Figure 3.3 Project scientists, water managers and a farmer near the Glomma River, Norway

Source: L. J. Barkved

Collaborative approaches to participation, which includes social learning, have recently attracted a high level of attention within this domain. Social learning is an important addition to the participatory armoury and is viewed as:

- appropriate within 'management regimes that require changes in social practices, roles and responsibilities' (Pahl-Wostl and Hare, 2004, p193);
- a complementary policy option when preferred outcomes are not clear and

subsequently cannot be based on fixed forms of knowledge (Collins and Ison, 2006).

Regarding whether and when social learning should be encouraged, Mostert et al (2007, pp 13–16) consider that social learning is nothing extraordinary as it happens:

> ... whenever interdependent stakeholders with different interest and perceptions come together and manage to deal with their differences to the benefit of all involved. Social learning really becomes an issue in complex organizational settings and in controversial cases in which it does not occur naturally. In these situations, social learning processes can become time consuming and costly and often require professional facilitation.

Drawing from Ridder et al (2005) they conclude that social learning processes should only be used when dealing with issues of central importance, or where there is a remote chance of success. Furthermore, where a social learning approach applies, policy-making in the traditional sense does not become immaterial but can be 'encompassed within a broader understanding of how knowledge, and thus issues, are constructed and employed in policy processes. A social learning approach provides a context for a dynamic local decentralized process, and, in the case of large watersheds, for concerted parallel local processes' (Collins and Ison, 2006).

While ongoing concerted and inclusive action can modify and transform situations, a good process does not necessarily lead to a good outcome (Creighton, 2005). Unrealistic assumptions are frequently made in relation to providing the public opportunity for both access to and incorporation of their contributions within the decision-making process, resulting in 'the effect of discouraging participation, encouraging conflict and fostering distrust among the participants' (Doelle and Sinclair, 2006, p187). The flow, content and relevance of a process to participants may also change over time. Individuals or organizations may either leave or join in during the course of a longer-term process, as and when the process necessitates their input, or due to a change in circumstances.

Ultimately, there is no 'one size fits all' solution to participatory processes, as different tools and approaches may be required throughout the lifetime of a particular water resource management (WRM) project. With these considerations in mind, it is recommended that:

> ... methods should therefore match the goals and purposes of participatory processes as much as the attributes of stakeholders and participants. Specifically, context characteristics can change in the course of a longer participatory process, such that it may be suitable to employ more formalized methods at certain stages and less formalised methods at other stages. (Newig et al, 2008, p386).

Similarly, Hartley (2003) stresses that it is not merely a matter of choosing which participatory tool to apply; rather, it is about employing and adapting as many as necessary to the needs of the stakeholders concerned.

Stakeholder identification and inclusion within case study basins

A number of key methods were used in relation to stakeholder participation in the four case study areas. A preliminary stakeholder analysis was conducted in order to identify the relevant stakeholders within each of the areas and to ascertain their interests, and the institutional and legal frameworks in which they operate. Initially this was accomplished by doing a survey together with the partner institutions located in the basins, as well as secondary sources. The analysis outlined the main institutions and stakeholder groups in the basins: government agencies, the private sector and civil society organizations such as consumer and agricultural associations. The analytical framework used in the project looked at governance using the specific parameters of how the institutional and legal/policy frameworks support stakeholder participation in water governance. Three methodologies comprised the analytical framework of the research: stakeholder analysis, legal analysis and actor network theory (ANT). Using these three methodologies the relative decision-making power of different stakeholder groups is described. Procedural rules providing access to information and decision-making were examined and in two basins the legal framework was examined through the prism of accountability and transparency using indicators to verify legal implementation in relation to water management and IWRM.

After looking at the various stakeholder groups, key institutions were identified and representatives from these institutions and NGOs were invited to the stakeholder workshops. Three workshops, one each year, were held in each basin except in Cambodia because the government partner did not participate in the project; here, field visits were conducted to look at the stakeholder situation. It was not always easy in each basin to obtain the participation of some of the main institutions, so further field visits were conducted in order to obtain information about the institutional framework and dynamics. These interviews filled in some of the gaps at the workshops.

At the workshops a number of common themes emerged, such as the impacts of industry, agriculture, dams, urbanization and climate change upon the river; water pricing; public education about water issues; and access to decision-making to participate in river basin planning, monitoring and management. After participants identified the main issues, they examined areas of conflict in the use of the water resources and the impacts of different water users upon the others – for instance, the impact of dams upon water flows and fishing communities in the Tungabhadra or the impact of large dams upon fishing communities living along the river. Another common theme that emerged was the need for public education about water and environmental issues. It was felt that the general public could be mobilized to do more in terms of water and biodiversity conservation if they understood the threats to the river. To this end the participants also mentioned the need for different stakeholders to be able to meet on a regular basis, and to have a multi-stakeholder committee that could help them to look at the issues and to find some solutions together. At the first Tungabhadra stakeholder workshop in Hospet, such a committee was formed. However, without funding and a project structure to support such initiatives, it will be difficult either to establish or to maintain such a committee. Expert councils and associations exist in the European basins, performing the function of such a committee to a degree; but these could be said to be more expert led, and specific interest groups could thus push through their agendas, rather than respond, in some cases, to broader governance and environmental issues. Scenarios – discussed further in the Chapter 4 – were also developed by the participants to look at how to deal with pressures on water resources such as rapid urbanization and climate change, as well as the impacts of dams upon communities and biodiversity.

In terms of the legal and policy framework, an examination of how the procedural rules regarding access to information and access to decision-making are implemented in the four basins was conducted. This allowed the research team to demonstrate the legal basis for stakeholder participation in IWRM. Implementation was monitored in the Tagus and the Sesan Basin through indicators developed regarding IWRM. Of course, the situation is ever changing in response to the pressures facing water managers, as well as the evolving participation process through conducting river basin plans under the Water Framework Directive in the Tagus; but the research revealed that implementation was not always as effective as it might be. In terms of access to information and decision-making, procedural rules are in place, often related to environmental impact assessments. Additionally, there is public participation in the development of river basin plans in the case of all basins except the Tungabhadra, which does not have a basin-wide plan. However, research shows the dominance and, in some cases, regulatory capture of hydropower and, in the case of the Tungabhadra, the agricultural sector. This was also reflected by the extent to which overall water resource management in the basins is connected to the administration responsible for the principal water uses. Water resources/ quantity management in Norway is broadly the responsibility of the Ministry of Petroleum and Energy, which is also responsible for hydropower; in India, water resources are managed by renamed departments of irrigation; and lower-level water management agencies in Vietnam are associated with irrigation and hydropower management. Access to information laws also allow stakeholders to access information through public authorities. But while a number of conventions and national laws and policy documents provide for stakeholder participation in water management, practice demonstrates that government and industry tend to dominate decision-making, so that interest groups representing environmental concerns and local communities tend to have less influence in decision-making despite the procedural rules allowing them access to relevant information and procedures such as public inquiries.

Conclusions

The existing body of knowledge and understanding on stakeholder participation in the context of water governance shows that, while the general concept has been embraced within the applicable laws and policies, more understanding is needed in order to comprehend what type of participation works well in which context. Such an understanding requires an appreciation of the interrelationship between the institutional, legal, political, historical and cultural environment present within a particular case area. In building upon the existing body of research, this collection of works has sought to further knowledge and understanding through a range of methods, including stakeholder analysis, actor network theory, governance assessment, workshops, focus groups and so forth. The evaluation of these methods is beyond the scope of this chapter, but will be a central component of the following chapters, which critically reflect upon participatory methods within the context of scenario development (Chapter 4), pollution assessment (Chapter 5), land and water interactions (Chapter 6), environmental flow (Chapter 7) and transboundary regimes (Chapter 8).

References

Abels, G. (2007) 'Citizen involvement in public policy making: Does it improve democratic legitimacy and accountability? The case of pTA', *Interdisciplinary Information Sciences*, vol 13, no1, pp 103–116

Abels, G. and Bora, A. (2004) *Demokratische Technikbewertung*, Transcript, Verlag, Bielefeld

Agenda 21 (1992) 'A Programme for Action for Sustainable Development', Rio de Janeiro, Brazil, June 13 1992, in *Report of the United Nations Conference on Environment and Development*, Annex II, UN Doc.A/Conf.151/26 (Vol. II)

Antunes, P., Kallis, G., Videira, N. and Santos, R. (2009) 'Participation and evaluation for sustainable river basin governance', *Ecological Economics* vol 68, no 4, pp931–939

Beierle, T. (1998) 'Public participation in environmental decisions: an evaluation framework using social goals'. Discussion paper 99–06, *Resources for the Future*, Washington DC.

Biswas, A. K. (2004) 'Integrated water resources management: a reassessment', *Water International*, vol 29, no 2, pp248–256

Bouwen, R. and Taillieu, T. (2004) 'Multi-party collaboration as social learning for interdependence: Developing relational knowing for sustainable natural resource management', *Journal of Community and Applied Social Psychology*, vol 14, pp137–153

Brady, K. (1998) 'New convention on access to information and public participation in environmental matters', *Environmental Policy and Law*, vol 28, pp69–75

Bruch, C. and Filbey, M. (2002) 'Emerging global norms of public involvement', in *The New 'Public': The Globalization of Public Participation*, The Environmental Law Institute, Washington, DC

Burgess, J., Clark, J. and Harrison, C. M. (2000) 'Knowledges in action: An actor network analysis of a wetland agri-environment scheme', *Ecological Economics*, vol 35, pp119–132

Cash, D., Clark, W., Alcock, F., Dickson, N., Eckley, N. and Jäger, J. (2002) *Salience, Credibility, Legitimacy and Boundaries: Linking Research, Assessment and Decision Making*, John F. Kennedy School of Government, Harvard University, Faculty Research Working Paper Series RWP02-046, US

Collins, K. and Ison, R. (2006) 'Dare we jump off Arnstein's ladder? Social learning as a new policy paradigm', Paper presented to the Participatory Approaches in Science and Technology Conference, Edinburgh, 4–7 June 2006

Creighton, J. L. (2005) 'What water managers need to know about public participation: One US practitioner's perspective', *Water Policy*, vol 7, pp269–278

Doelle, M. and Sinclair, S. A. (2006) 'Time for a new approach to public participation in EA: Promoting cooperation and consensus for sustainability', *Environmental Impact Assessment Review*, vol 26, no 2, pp185–205

du Toit, D. and Pollard, S. (2008) 'Updating public participation in IWRM: A proposal for a focused and structured engagement with Catchment Management Strategies', *Water SA*, vol 4, no 6, pp707–714, www.wrc.org.za, last accessed March 2010

Dublin (1992) *Statement on Water and Sustainable Development*, Dublin, Ireland, 31 January 1992 (reprinted *in Environmental Policy and Law*, vol 22, pp54–60)

EC (2003a) *Communication on Governance and Development*, Com (03) 615, EC, Luxembourg

EC (2003b) *Common Implementation Strategy for the Water Framework Directive (2000/60/EC)*, Guidance Document No 8, Public Participation in Relation to the Water Framework Directive, EC, Luxembourg

Gooch, G. D. (2004) 'The communication of scientific information in institutional contexts: The specific case of transboundary water management in Europe', in J. G. Timmerman and S. Langaas (eds) *Environmental Information in European Transboundary Water Management*, IWA Publishing, London

Gooch, G. D. and Huitema, D. J. (2008) 'Participation in water management: Theory and practice', in J. G. Timmerman, C. Pahl-Wostl and J. Möltgen (eds) *The Adaptiveness of IWRM – Analysing European IWRM Research*, IWA Publishing, London

Gray, B. (1989) *Collaborating: Finding Common Ground for Multiparty Problems*, Jossey Bass, San Francisco, CA

Gray, B. (2004) 'Strong opposition: Frame-based resistance to collaboration', *Journal of Community and Applied Social Psychology*, vol 14, pp166–176

Hague (2000) *Ministerial Declaration of the Hague on Water Security in the 21st Century*, The Hague, The Netherlands, 22 March 2000, www.worldwaterforum.org, last accessed March 2010

Hare, M., Letcher, R. A. and Jakeman, A. J. (2004) 'Participatory modelling in natural resource management: A comparison of four case studies', *Integrated Assessment*, vol 4, no 2, pp62–72

Hartley, N. and Wood, C. (2005) 'Public participation in environmental impact assessment – implementing the Aarhus Convention', *Environmental Impact Assessment Review*, vol 25, no 4, pp319–340

Hartley, T. W. (2003) *Water Reuse: Understanding Public Perception and Participation*, Water Environment Research Foundation, US.

Heere, W. P. (2004) *From Government to Governance*, Cambridge University Press, Cambridge

Howarth, B. (2007) 'Substance and procedure under the strategic environmental assessment directive and the water framework directive', in J. Holder, and D. McGillivray (eds) *Taking Stock of Environmental Assessment – Law, Policy and Practice*, Taylor and Francis, Abingdon, UK, pp149–190

Huitema, D. (2005) 'A comparative analysis of the three citizens' juries under River Dialogue: Focus groups and citizens' juries', in K. Kangur (ed) *River Dialogue Experiences in Enhancing Public Participation in Water Management*, Peipsi Center for Transboundary Cooperation, Tartu, Estonia, pp44–50

Janssen, M. A., Goosen, H. and Omtzigt, N. (2006) 'A simple mediation and negotiation support tool for water management in the Netherlands', *Landscape and Urban Planning*, vol 78, no 1, pp71–84

Johannesburg (2002) 'Declaration of the World Summit on Sustainable Development', in *Report of World Summit on Sustainable Development*, Johannesburg, South Africa, 26 August–4 September 2002, UN Doc. A/Conf. 199/20 (2002)

Kallis, G., Vidiera, N., Antunes, P. and Santos, R. (eds) (2004) 'Integrated deliberative decision processes for water resources planning and management – a guidance document, report prepared for the ADVISOR project (under contract EVK1-CT2000-00074), http://coman.dcea.fct.unl.pt/projects/advisor/publications.htm

Loucks, D. P. (2006) 'Modeling and managing the interactions between hydrology, ecology and economics', *Journal of Hydrology*, vol 328, nos 3–4, pp408–416

Mitchell, R. K., Agle, B. R. and Wood, D. J. (1997) 'Toward a theory of stakeholder identification and salience: Defining the principle of who and what really counts', *Academy of Management Review*, vol 22, no 4, pp853–886

Mostert, E. (2003) 'The challenge of public participation', *Water Policy*, vol 5, pp179–197

Mostert, E., Pahl-Wostl, C., Rees, Y., Searle, B., Tàbara, D. and Tippett, J. (2007) 'Social learning in European river-basin management: Barriers and fostering mechanisms from 10 river basins', *Ecology and Society*, vol 12, no 1, p19, www.ecologyandsociety.org/vol12/iss1/art19/, last accessed March 2010

Newig, J., Pahl-Wostl, C. and Sigel, K. (2005) 'The role of public participation in managing uncertainty in the implementation of the Water Framework Directive', *European Environment*, vol 15, no 6, pp333–343

Newig, J., Haberl, H., Pahl-Wostl, C. and Rothman, D. S. (2008) 'Formalised and non-formalised methods in resource management – knowledge and social learning in participatory processes: an introduction', *Systemic Practice and Action Research*, vol 21, pp381–387

Özerol, G. and Newig, J. (2008) 'Evaluating the success of publication participation in water resources management: five key constituents', *Water Policy*, vol 10, no 6, pp639–655

Pahl-Wostl, C. (2008) 'The importance of social learning and culture for sustainable water management', *Ecological Economics*, vol 64, pp484–495

Pahl-Wostl, C. and Hare, M. (2004) 'Processes of social learning in integrated resources management', *Journal of Community and Applied Social Psychology*, vol 14, pp193–206

Pahl-Wostl, C., Craps, M., Dewulf, A., Mostert, E., Tabara, D. and Taillieu, T. (2007) 'Social learning and water resource management', *Ecology and Society*, vol 12, no 2, www.ecologyandsociety.org/vol12/iss2/art5/, last accessed March 2010

Phillips, D. J. H., Daoudy, M., Öjendal, J., Turton, A. and McCaffrey, S. (2006) *Transboundary Water Cooperation as a Tool for Conflict Prevention and for Broader Benefit-sharing*, Ministry for Foreign Affairs, Stockholm, Sweden

Priscolli, J. D. (2003) *Participation, Consensus Building and Conflict Management Training Course*, UNESCO/IHP/WWAP/IHP-VI Technical Documents in Hydrology PCCP Series no 22, UNESCO, France

Rault, K. P. and Jeffrey, P. (2008) 'Deconstructing public participation in the Water Framework Directive: Implementation and compliance with the letter or with the spirit of the law?', *Water and Environment Journal*, vol 22, pp241–249

Renn, O. (2008) *Risk Governance: Coping with Uncertainty in a Complex World*, Earthscan, London

Ridder, D., Mostert, E. and Wolters, H. A. (eds) (2005) *Learning Together to Manage Together: Improving Participation in Water Management*, University of Osnabrück, Osnabrück, Germany, www.harmonicop.vos.de/HarmoniCOPHandbook.pdf, last accessed March 2010

Rogers, P. and Hall, A. W. (2003) *Effective Water Governance*, GWP, Stockholm

Stec, S. and Casey-Lefkowitz, S. (2000) *The Aarhus Convention: An Implementation Guide*, United Nations, Geneva

Sharp, L. (2002) 'Public participation and policy: Unpacking connections in one UK Local Agenda 21', *Local Environment*, vol 7, no 1, pp7–22

SLIM Project (2004) 'Introduction to SLIM publications for policy makers and practitioners', www.transboundarywater.se/documents/SLIM/Slim%20introduction.pdf, last accessed March 2010

Stec, S. and Casey-Lefkowitz, S. (2000) *The Aarhus Convention: An Implementation Guide*, United Nations, Geneva

Steyaert, P. and Jiggins, J. (2007) 'Governance of complex environmental situations through social learning: a synthesis of SLIM's lessons for research, policy and practice', *Environmental Science and Policy*, vol 10, pp575–586

Tilleman, W. A. (1995) 'Public participation in the environmental impact process: A comparative study of impact assessment in Canada, the United States and the European Community', *Columbia Journal of Transnational Law*, vol 33, pp337–359

UNDP (United Nations Development Programme) (2004) *Strategy Note on Governance for Human Development*, UNDP, New York, NY

UNECE (United Nations Economic Commission for Europe) (1998) *Convention on Access to Information, Public Participation in Decision-Making and Access to Justice in Environmental Matters*, 25 June 1998 (entered into force 30 October 2001), reprinted in 38 I.L.M. 517 (1999), www.unece.org/env/eia/sea_protocol.htm, last accessed March 2010

UNECE (2003) *Protocol on Strategic Environmental Assessment*, 21 May 2003 (not yet in force)

UNWDR (2006) Water: A Shared Responsibility – UN World Water Development Report, UNESCO, Paris

van Eeten, M. J. G., Loucks, D. P. and Hermans, L. (2003) 'Bringing actors together around large-scale water systems: Participatory modeling and other innovations', *Technology and Policy*, vol 14, no 4, pp94–108

Vega-Leinert, A. and Schröter, D. (2008) 'Stakeholder dialogue as a communication and negotiation tool in scientific inquiry', in A. Carvalho (ed) *Communicating Climate Change: Discourses, Mediations and Perceptions*, Centro de Estudos de Comunicação e Sociedade, Universidade do Minho, Braga, www.lasics.uminho.pt/ojs/index.php/climate_change, last accessed March 2010

Videira, N., Antunes, P., Santos, R. and Lobo, G. (2006) 'Public and stakeholder participation in European water policy: A critical review of project evaluation processes', *European Environment*, vol 16, pp19–31

Welp, M., Vega-Leinerta, A., Stoll-Kleemann, S. and Jaeger, C. C. (2006) 'Science-based stakeholder dialogues: Theories and tools', *Global Environmental Change*, vol 16, pp170–181

Wester, P., Merrey, D. J. and de Lange, M. (2003) 'Boundaries of consent: Stakeholder representation in river basin management in Mexico and South Africa', *World Development*, vol 31, no 5, pp797–812

4

The Science–Policy–Stakeholder Interface in Sustainable Water Management: Creating Interactive Participatory Scenarios together with Stakeholders

Geoffrey D. Gooch, Andrew Allan, Alistair Rieu-Clarke and Susan Baggett

Introduction

As has been noted in earlier chapters in this book, a major problem in the move to sustainable water management is the incorporation of scientific and other forms of information (such as local knowledge) within the science–policy interface and within the policy process itself. Despite the considerable amount of effort put into research on the factors influencing policy input, the results of this knowledge production are often not successfully incorporated within policy-making. The reasons for this are many (see, for example, Chapter 2) and (in cases) uncertain. Part of the problem seems to be in the initial formulation of problems; scientists are concerned with problem-solving procedures that fit into scientific disciplines or that can be judged by others in the scientific community (see Chapter 2). Policy-makers and managers, on the other hand, need answers to more immediate problems. The correlation between these two ways of looking at problems is often weak. At the same time, stakeholders can provide an important source of knowledge for water management, a knowledge that can both complement and further inform the knowledge provided by the scientific community. A possible solution to this is that scientists could make their methods and results more understandable and accessible to policy-makers, while policy-makers could also make clearer the type of information necessary for policy formulation and implementation. Into this two-way exchange we need to insert a third: the knowledge produced by local people and stakeholders. As will be seen later in Chapter 7, local knowledge is often a vital input into policy, especially in scientifically data-poor contexts, such as our examples from South-East Asia and India.

As shown in Chapter 3, the increased interest and demands on stakeholder and public participation have highlighted these issues. It is now generally considered vital, also from a legal perspective, to include these groups in water management processes (see Chapter 3). How, then, to include stakeholders and the public in *informed* policy-making in such a technically complicated field such as water management? Participation can be used not only as a way to represent interests, but also as a way to provide input into the policy process, by which we mean that informed participation is based on knowledge, either local or/and scientific. A common question asked is: are stakeholders and the public at all able to understand the complexity of the issues at stake? If not, how can they participate in an informed manner? Based on the results of earlier research (Gooch, 2004), the answer to this question is: yes, under certain conditions it is perfectly possible for these groups to both understand and contribute to policy for sustainable water management. Nowadays there are a wide range of tools that enable stakeholders to participate in environmental management (Gooch and Huitema, 2008) such as round tables, citizen juries, panels, etc., and the choice of these is dependent on the context and aims of the participatory exercise (see Chapter 3). However, while these methods and tools may enable laymen to comprehend complex problems and to contribute to the policy process, they do not necessarily provide a means of involving non-experts in planning for the future. On the other hand, who exactly *are* the experts who can discern the future? Here the problem is extenuated through a lack of information about future conditions; we are faced with conditions of uncertainty. At the same time, the scope of issues concerning sustainability is so wide that they encompass all aspects of society; therefore, all parts of society should be included in formulating possible solutions. One way of involving stakeholders and the public in the formulation of possible futures is through the use of scenarios – projections of possible futures (Alcamo, 2001; Shell, 2003).

As a way of improving the science–policy–stakeholder interface (SPSI) in our case basins, particularly the interaction between stakeholders and scientists, interactive participatory scenarios were developed within the four case study areas. Stakeholders and the public were especially included in the formulation of scenarios for sustainable water management on the Sesan (Vietnam–Cambodia) and Tagus (Spain–Portugal) rivers. These scenarios were constructed in a three-step process, through which scientists first conducted analyses of management regimes, physical conditions, existing agreements, etc., and then formulated preliminary scenarios that were presented to stakeholders at the first series of workshops. Water management regimes were first examined by looking at actor networks, communication processes, law and governance issues. We saw that actor networks in all the case areas consisted of formal and informal actors and institutions, as well as non-human entities such as dams and irrigation systems. Communication processes were of central importance as it is within these that information, knowledge and mutual understanding of problems and their solution are formed. Law plays a vital role in conferring rights and obligations on actors in support of integrated water resource management (IWRM) and existing agreements and conventions were studied to provide an understanding of legal frameworks. However, the formal adoption of appropriate laws is meaningless without securing their effective implementation, and for this it is necessary to secure the support of stakeholders and the public. Qualitative interactive and participatory scenarios were therefore used as a means of involving stakeholders in the formulation of policies, as well as a way of improving social learning processes and the potential of policy implementation. These processes are described in the following sections.

Scenarios describe plausible futures. However, the utility of both the process and the products are also vested in the present and can help to identify contemporary problems and issues. Knowledge about key drivers and uncertainties of the future can provide information towards better decisions in the present and a better understanding of key drivers and the possible trajectories of changes made. In turn, this collection of knowledge and provision of information will not only clarify the impact of decisions, but may also facilitate active countering of undesirable trajectories of change. The identification and characterization of key uncertainties can enable a more structured approach to risk management. Strategies and decisions can be played out in different futures to secure the most beneficial outcome through the most robust approaches with the least risk. The process can produce new knowledge that not only benefits resource managers and decision-makers, but also empowers stakeholders. This extends from politicians and policy-makers through to government officials, the private sector and civil society. Knowledge generated and sourced through structured participative research processes can increase understanding, both through participating in the process and by accessing appropriate forms of communication. The identification, description and ranking of key drivers and uncertainties can be translated to potential implications. Scenarios have been the focus of increased interest and initiatives such as those published by the European Environmental Agency on the Environmental Scenarios Information Portal (http://scenarios.ew.eea.europa.eu/), which have contributed to the development of this field. Yet, the use of scenarios as a means of improving stakeholder participation and the science-policy interface still needs to be developed.

Science–policy interface

A major problem for policy-makers is, as has been noted above, obtaining reliable and relevant information upon which to base decisions. This information may be provided by the scientific community, but may also come from other sources. For example, it is vital for policy-makers to be aware of public opinion, as political leadership in democracies is based on citizen support. This type of information can be provided by opinion polls, through direct communication with the public, through civil society organizations and NGOs, and through the mass media. Scientific information must therefore compete with other forms in an information marketplace. This is rarely understood by the scientific community who often believe (incorrectly) that their form of information speaks for itself. We have also claimed that policy-makers need both reliable *and* relevant information. By this we mean that information often needs to be user driven as opposed to being the result of disciplinary and scientific priorities. Basic research is, of course, vital, as all knowledge cannot be user driven; but a high proportion of scientific output never enters the policy process. This is because it is not considered relevant, because it is not understood, or because it does not fit into other priorities, such as gaining and keeping public support. In order to increase scientific input it is therefore necessary to tailor research to policy needs, and/or to improve the communication of scientific results. By including stakeholders and, in some cases, local and regional policy-makers in the process of scenario construction, we were able to present scientific knowledge in a form that was considered both relevant and understandable for the policy-makers. In the cases where policy-makers participated, the process also helped them to define future research objectives, as the scenarios helped to identify future challenges. Furthermore, the inclusion of legal and institutional concerns in the preparation of the scenarios was designed to assist policy-makers in understanding the importance of the role of governance in implementing policy priorities.

Scenarios in the science–policy interface

Scenarios have been used increasingly during recent years, the most high-profile global examples being seen in the context of the IPCC's assessment reports (Nakicenovic et al, 2000). They have also been used more specifically in the context of water and basin management (e.g. Seckler et al, 1998; Gallopin and Rijsberman, 2000; Hope, 2006; Kämäri et al, 2008), with use being made of both quantitative (Pallottino et al, 2005) and qualitative (e.g. Gallopin, 2006) scenarios, and combinations of both (Gooch and Stålnacke, 2006).

In this latter category, the SCENES project makes use of multifactor integrated scenarios with respect to the European position, in particular (Kämäri et al, 2008), using both pre-existing scenarios and iterative processes for the development of new ones to quantitatively bolster the storylines. Gallopin and Rijsberman (2000) have examined scenarios in the global context, following the World Water Vision, and have derived conclusions that could be alarming for policy-makers – principally that qualitative differences in scenarios over a relatively short time period of 25 years can, in fact, be substantial (Gallopin, 2006), with the corollary that action taken in the very short term can have significant effects over the medium. The importance of the impact of actions in other separate but connected sectors was also emphasized, highlighting the need for integrated scenarios. In a more quantitative approach, the IWMI has applied scenarios to a single sector, irrigation (Seckler et al, 1998), in order to give flesh to their model and show how it works in practice. Normal practice is to include a 'business as usual' scenario among the proposals (Seckler et al, 1998; Kämäri et al, 2008), which functions largely as a benchmark against which other futures can be measured, with four scenarios being generally held to be the optimum number (Nakicenovic et al, 2000; Peterson et al, 2003). Furthermore, as in the case of the MANTRA East project, qualitative scenarios are not always viewed as an end in themselves, but are used instead as an input into the quantitative scenarios, thereby forming an integrated process utilizing both qualitative and quantitative approaches (Gooch and Stålnacke, 2006). Scenario development has the potential to become a dominant tool for engaging a diverse set of stakeholders; not only can it be developed without the need for a high degree of technical ability, but it can also be easily understood by an extensive set of actors and has the flexibility to incorporate in-depth knowledge from a broad set of disciplines (Kok et al, 2007).

The degree to which stakeholders are involved in the development of scenarios differs between authors, projects and the type of scenario being used. With respect to the types of scenario available, it is clear that a variety of typologies exist for differentiating these. Börjeson et al (2006) identify three broad types of scenario – predictive, explorative and normative – all of which incorporate three broad stages, from the generation of ideas and data, through to integration and then to consistency checking. They highlight the relative importance in each of surveys, interviews and the utilization of techniques such as Delphi methods, highlighting the different ways in which the views of stakeholders can be incorporated within scenario-building. Van Notten et al (2003), on the other hand, take the view that classifications become outdated as the science develops, and consequently recommend a typology based on an analysis of how scenarios develop – namely, project goal, process design and scenario content, with the spectrum of stakeholder involvement being one aspect of this, in terms of the method of data collection.

In practice, the extent to which stakeholders have been involved in the development of water-related scenarios has varied. The results of the SIRCH project in England, which combined quantitative models and storylines, highlighted the fact that participatory scenarios are more useful in terms of eliciting greater insight from stakeholders than predetermined ones, in the view of Pahl-Wostl (2002). Gallopin and Rijsberman (2000) make use of the scenarios that were developed through the World Water Vision process, and these highly detailed storylines were the product of an iterative dialogue with a large number of stakeholders who were involved with the development, refinement and validation of these products (Gallopin, 2006). The IWMI process, unlike the others mentioned here, does not make use of stakeholder workshops to elucidate the key drivers and uncertainties; but its scenarios have a rather more specialized function to enhance their model rather than using the scenarios as mechanisms for enabling knowledge transfer to policy-makers. Kämäri and his group use a series of stakeholder workshops to develop and elaborate storylines, with the question of geographical scale being key (Kämäri et al, 2008). Groves and Lempert (2007) propose a methodology for reducing the infinite number of potential futures to three or four scenarios that are policy relevant: it seeks to address the perceived failings of the scenario-axes methods used in most other studies, where scenarios are merely reflections of the concerns of authors and where probabilities are ignored. They transmit a view indicating that a robust decision-making tool, coupled with algorithms, is capable of taking policy actions as its basis, and generating scenarios that relate directly to the policy-making world from a quantitative perspective. At the end of the process, the role of scenarios as a policy development tool becomes critical, given their potential for packaging future uncertainty into parcels that can be more easily processed by policy-makers (EEA, 2009). This is particularly the case with respect to water, which falls into the category of deep uncertainty, as used by Groves and Lempert (2007). Making sense of this uncertainty is clearly foremost in their minds as their work focuses on this policy implementation aspect, and they are critical of the traditional stakeholder-led process of scenario development insofar as it does not directly link to the potential for remedial action alternatives (Groves and Lempert, 2007). The SCENES methodology implicitly accepts these criticisms of the traditional approach, and instead sets out its own system that aims to ensure a more effective science–policy interface (Kämäri et al, 2008). SCENES intends to make scenarios more policy relevant through the involvement of decision-makers at the development stage, and a specific undertaking to sit within current management and planning practices. By careful selection of stakeholders and placement of these participants in scales assessing their relative importance in decision-making (both current and future), they presumably hope to ensure maximum relevance, and evaluation of the scenario-building process will include a component relating to the extent to which stakeholders have accepted the final 'possible futures'.

Actor networks in water management scenarios

One of the first steps in formulating the scenarios in our case basins was the identification of actor networks, partly because these networks influence possible futures, and partly because we needed to identify potential end users for our scenarios. Actor networks include formal actors such as government departments at national, regional and local levels, as well as NGOs, local people and other stakeholders. These actors work within networks formed by national and international laws and by accepted practices, the 'rules of the game'. While it has been common to try to distinguish between different spatial levels in water management regimes, this seems now to be somewhat outdated, as actor networks can stretch from local to international levels through the activities of organizations such as the United Nations, aid agencies and international NGOs. In the case of sustainable water management, these are especially important, as it is through actor networks that information can be exchanged and future developments discussed. In order to facilitate positive developments in sustainable water management, all relevant actors should be included in the definition of what the network is and should be. Actor networks in water management include organizations, stakeholders' institutions, and non-human entities, such as hydroelectric power dams and village meeting houses. All should be taken into account in order to understand the forces influencing sustainable water management (see Chapter 8).

Communication in water management scenarios

A second step in the process of our scenario-building was the identification of channels of communication. Decisions in water management are based on a combination of information and perceptions of what the main problems are and how best to solve them. In both cases it is necessary to analyse ways of communication between actors and how, if at all, basic information about water flows, quality, etc. is exchanged. How, if at all, are decisions concerning socio-economic developments arrived at? Information exchange may take place through contacts between organizations, common databases or through the facilitation of third parties such as international organizations. Communication with stakeholders and the public is especially important, and in areas with low levels of literacy, as in the cases presented here, it is often only achieved through personal communication, group discussions and through graphic representations such as information posters. For example, in the Sesan Basin in Cambodia, posters describing fish-farming techniques were used by the Cambodian fisheries department. These fish-farming techniques represented a possible future development for the basin, in which changes in water flow and overfishing have seriously depleted fish stocks, particularly of migratory fish species. Non-verbal information such as this can be an important communication method in places with low literacy levels as there is no point in providing long technical reports to people without the necessary reading skills. Communication also takes place between and within water management institutions, as well as with stakeholders and the public, and the means of communication needs to be adjusted to these different groups and goals. It is also important to remember that communication is a two-way process, while information can move in one or more directions. Local knowledge and insights enrich the water policy process, and by involving stakeholders and the public in the process, stakeholders can also develop their communication skills and gain other forms of knowledge.

Legal and institutional aspects of the scenarios

In order to create the legal and institutional contexts for future scenarios for the case basins, it was necessary to first identify and analyse the role of law and administration. Law plays a vital role in conferring rights and obligations on actors, which, in turn, can provide a legitimate and predictable framework by which to support sustainable water management. There are, in effect, two levels at which law has to work. First, it must contain appropriate provisions drafted to permit effective interpretation and implementation. Second, however, simply putting such laws in place is of little worth if they are not supported by effective implementation systems. While the degree to which effective implementation is possible depends on financial, human resource, administrative and infrastructure capacity, it is also critically reliant on the existence of a broader governance framework that supports the rights of access to justice, information and participation in decision-making. These governance frameworks are an important part of our water management scenarios. Additionally, given the indivisibility of river basins, rights and obligations must be in place to secure a basin-wide approach to water resources management. National water laws and transboundary basin agreements must therefore incorporate rules and principles that recognize the linkages between land and water irrespective of political boundaries and that facilitate coordination between the various vertical decision-making levels, where responsibilities for land, water and related sectors are split between agencies. This legal framework must also ensure that surface and groundwater are integrated within the allocation processes, and that equitable access for all is assured.

The broader governance framework referred to above in relation to access to justice, access to information and participation in decision-making plays a further key role in the incorporation of stakeholder views and research within policy development. If sustainable use of water resources is to be achieved, it is imperative, as noted above, that it takes account of the experience of stakeholders and relevant actor networks, along with the best relevant research from the natural and social sciences. The governance framework will establish the formal mechanisms through which this knowledge feeds into the policy development process in as transparent and accountable a way as possible. The legal analysis of our case study basins therefore tried to take into account not only the legal framework surrounding the management of water and land resources, but also the wider administrative law context, constitutional law and the efficiency of the administration of justice. Water management takes place within this wider context of law and administration, and cannot be understood correctly without knowledge of this wider context.

The legal analysis of the river basins showed that while considerable efforts had been made in adopting the appropriate laws in support of sustainable water management in most places, implementing such laws still remains a challenge. In particular, there was a need to strengthen stakeholder participation and the existing institutions in order to support a true basin approach to water resources management.

Interactive participatory scenarios as a policy tool

This brings us to a central issue: how interactive participatory scenarios can be used as a means of both improving policy strategies and of involving stakeholders and the public in that policy-making. Scenarios present a way for the policy-maker, manager or scientist to test ideas about possible futures through exercises that can clarify the probable results of certain courses of action and implementation of particular governance frameworks. In our case basins they were also used as a means of improving public and stakeholder participation. As these groups were involved in the formulation and evaluation of scenarios, they

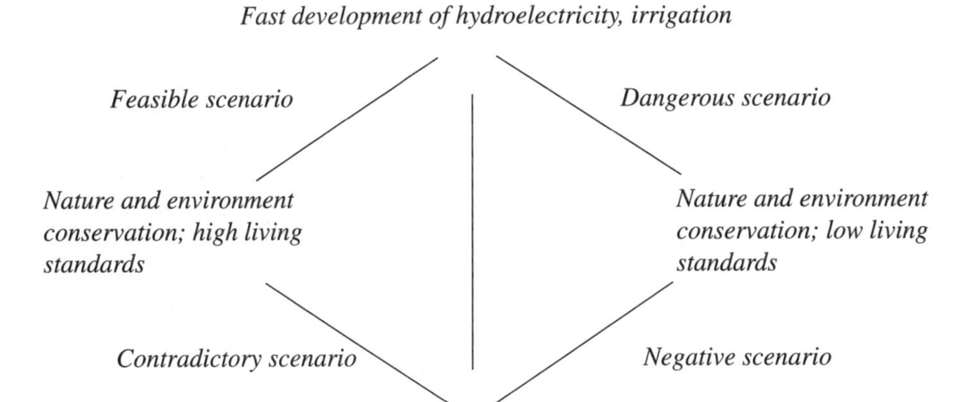

Figure 4.1 Preliminary scenarios used in the first stakeholder workshops

Source: G. D. Gooch

could provide insights not readily available for scientists and policy-makers, and the method could also increase the level of social learning and reduce the scope for ineffective resource allocation (Hope, 2006).

Within the case study areas, scenarios were formulated in partnerships with scientists, policy-makers and stakeholders. Since the scenarios were mostly developed in two of the case basins, the Tagus (Spain and Portugal) and the Sesan (Vietnam and Cambodia), the following descriptions of methodology will be mainly limited to these two basins. A number of phases were central to developing the scenarios, including stakeholder mapping, focus groups, actor network theory (ANT) analysis and the stakeholder workshops (see Figure 4.1).

Prior to the commencement of stakeholder workshops, stakeholder mapping exercises were conducted in order to identify the relevant stakeholder groups within each case study area. These included user groups such as water-user associations, community-based organizations, farmer associations, ethnic minority groups and so forth, along with relevant decision-taking institutions, policy-makers and representatives from the scientific community. The mapping exercise also sought to categorize existing knowledge of the interests of the stakeholders and institutions. The ultimate aim of the stakeholder mapping was to identify who should be engaged within the formulation of the scenarios and to what degree.

First rounds of workshops were then conducted within the case study areas based on the information from the stakeholder mapping exercise. The first workshop for the Tagus workshop took place in Toledo, Spain, on 14 December 2006, and brought together a range of actors from the scientific, stakeholder and policy communities in both Spain and Portugal. In the Sesan River Basin an initial stakeholder workshop took place in Pleiku, Vietnam, on 14 December 2006. The Pleiku workshop was only successful in bringing together stakeholders from the Vietnamese part of the Sesan, which included representatives from the scientific, policy and stakeholder communities, although an additional field trip to Cambodia was organized in April 2007. The central aim of the initial stakeholder workshops was to review the stakeholder mapping conducted by the research team, as well as to identify the key issues relating to the integrated management of the river basins. The initial workshops were supplemented by a series of focus groups and interviews within the case study areas; this provided the opportunity to interact with a wider range of stakeholders and policy-makers than was possible in the workshops. The meeting participants then identified who they saw as the main actors and what they regarded as being the key problems, taking into account such factors as population, governance, resource availability, politics and economics, among others. In addition, the identification of actors and their interests was supported through the ANT analysis, which is described in more detail in Chapter 8. This first phase of interaction between the research group, scientists, stakeholders and policy-makers within the case study areas was fundamental in providing the basis by which the scenarios could be formulated.

A second round of workshops took place for the Tagus in Lisbon, Portugal, on 28 February 2008, and for the Sesan in KonTum town, Vietnam, on 14 December 2007. Based on the issues identified during the first phase of workshops, focus groups and interviews, the research team was able to formulate four draft outline scenarios, which drew upon the issues identified by the various actors, (see Figure 4.2), but avoided the problems associated with the use of fewer or greater numbers of scenarios alluded to above. These draft scenarios were then presented at the second round of workshops as a basis for discussion and further elaboration. The first part of the discussion centred around developing a consensus upon which variables should be contained on the horizontal and vertical axes (see Figure 4.1). The axes reflected the main drivers within the basin, which for the Sesan were closely aligned to water and energy demands, as well as cooperation at the state-state level. Within the Tagus, international cooperation was also a key driver, along with the need to reconcile agricultural demands, on the one hand, and urban and tourism uses, on the other. Once consensus had been established with respect to the key drivers for the horizontal and vertical axes, the workshop participants were able to formulate four alternative scenarios for each of the river basins. In the context of the Tagus, these scenarios were business as usual; joint development for agriculture; conflicts over urban water supply; and joint development for tourism. For the Sesan, the four scenarios consisted of business as usual; joint hydropower development; joint development; and development of farming/fishing upstream in Vietnam. The outline scenarios were further refined as part of consultations held during more intensive field trips to Spain, Portugal, Vietnam and Cambodia.

The final rounds of workshops took place in Madrid, Spain, with respect to the Tagus, and in Vientiane, Laos, for the Sesan, in April and December of 2009, respectively. More refined scenarios were presented to the workshop participants for validation. These scenarios, which had been developed by the research team, provided further detail on the key social, economic, environmental and governance aspects of each of the four outline scenarios.

Through this process, stakeholder perspectives, governance analyses and the results of scientific research in the river basins were combined. The stakeholder workshops also helped to identify central aspects of the project, such as the actor networks. Finally, the development of scenarios highlighted particular aspects of the legal regime that may need to be altered and those aspects of the existing regimes that obstruct, affect or are likely to lead to particular scenarios, or that have an effect on the extent to which particular stakeholder groups can influence resource management.

While the physical, economic and social characteristics of the Sesan and Tagus basins differ significantly, the scenarios adopted by stakeholders in both were remarkably consistent in some respects, with the importance of trans-boundary cooperation in sustainable water management being recognized as central. Stakeholders in both basins regarded cooperation at this level as a core component of the application of IWRM, with the goal of sustainable development being uppermost in the stakeholders' priorities. They also took the view, however, that previous efforts aimed at enhancing inter-state basin management in their areas had been inadequate and that future development activity demanded more substantive management integration between riparian states. In the Sesan, the dominating role of hydropower was recognized by stakeholders; but the futures envisaged by them presented a spectrum of priorities, where hydropower is balanced to a greater or lesser extent with social and environmental concerns in a broader basin-wide context. Although no formal ranking of the four scenarios from the perspective of desirability was completed, a clear preference was expressed for the fourth scenario, 'multi-sector development and strong international cooperation', although there were some reservations regarding the feasibility of this 'future'. Although these scenarios reached their final stage of development in the third stakeholder meetings, it would have been beneficial to have further built on this process by formally allowing policy-makers to define more precisely their future research objectives (i.e. in effect, the policy optimization step suggested by Liu et al, 2007). Timing did not allow this; but the involvement of policy-makers in the scenario development groups ensured that they would be able to do this independently of the project. With respect to the role of law and wider governance concerns, it would also have been beneficial for further work to have taken place to align policy optimization with relevant implementation tools. This could then have provided policy-makers with road maps for the efficient application of policy, using cost-effective tools that capitalized on a clear appreciation of the existing governance position.

Figure 4.2 Stakeholders at the final Vietnamese–Cambodian workshop in Vientiane, Laos

Source: G. D. Gooch

Recommendations for the use of scenarios as a tool to involve stakeholders and the public

The use of scenarios showed that they can be a useful tool in involving stakeholders and improving the legitimacy and quality of policy strategies. Since the meetings were held in all the respective riparian states of the Tagus and Sesan rivers, stakeholders were able to formulate national perspectives, which were then combined at a meeting involving all participants. In the case of the Sesan River, this meeting, held at the Mekong River Commission offices in Vientiane in December 2008, was in itself an achievement: a problem that has complicated the formulation of policies for the sustainable use of the Sesan River has been a relative lack of communication between the Vietnamese and Cambodian stakeholders.

The major challenges to the use of scenarios are as follows:

- The process is time consuming. In order to succeed it is necessary to organize at least three or, preferably, four meetings: one to introduce the idea of scenarios, which can be difficult to grasp at once; a second meeting to discuss the first draft versions; a third to finalize the scenarios; and a fourth to enable policy-makers to define their policy objectives and implementation strategies in the light of the agreed scenarios.

- The selection of participants for the meetings is a central issue. It is, at the same time, necessary to create a group that can engage in creative discussions without excluding variant perspectives. This is important: while diametrically opposing perspectives may make the process more difficult, they are necessary if the specific function of scenarios, that of formulating possible futures (not necessarily probable), is to be achieved.
- In rural areas, especially in developing countries, the logistics of gathering people from a wider geographical area can be substantial. In the case of the Sesan River, for example, it proved difficult to arrange a meeting at the local level in the river basin because of unreliable and often unpredictable communications.
- The choice of factors upon which to base the scenarios is difficult. Simplification is necessary; but it is a very challenging exercise to strike a balance between accurate distillation and oversimplification of the complicated and inter-dependent issues related to sustainable water management.
- As with all processes that involve stakeholder participation, it is necessary to define realistic outcomes at the very beginning of the process. The results of an exercise such as this with scenarios can increase the understanding of the participants for common problems, can raise their awareness and knowledge, and can increase their capacity to define their own futures. However, it is necessary to make clear that the impact upon policy made by local, regional and national authorities is completely dependent on those authorities' willingness to take on the results of the process.
- Where possible, the scenario development process should not only focus on policy relevance, but also on the tools required by policy-makers to translate policy goals into reality effectively. An appreciation of existing legal and institutional frameworks is critical for this, and these played an important role in both defining the key basin issues and illuminating potential policy directions.

Despite these reservations, the use of stakeholder-formulated scenarios in the Sesan and Tagus basins have demonstrated that the method can be a useful contribution to the development of sustainable policies for water management.

References

Alcamo, J. (2001) *Scenarios as Tools for International Environmental Assessments*, European Environment Agency, Copenhagen

Börjeson, L., Höjer, M., Dreborg, K.-H., Ekvall, T. and Finnveden, G. (2006) 'Scenario types and techniques: Towards a user's guide', *Journal of Futures*, vol 38, no 7, pp723–739

EEA (European Environment Agency) (2009) *Looking Back on Looking Forward: A Review of Evaluative Scenario Literature*, European Environment Agency, Copenhagen

Gallopin, C. C. (2006) 'Linkages between vulnerability, resilience, and adaptive capacity', *Global Environmental Change*, vol 16, pp 293–303

Gallopin, G. C. and Rijsberman, F. (2000) 'Three global water scenarios', *International Journal of Water*, vol 1, no 1, pp16–40

Gooch, G. D. (2004) 'Improving governance through deliberative democracy – initiating informed public participation', Paper presented to Water Governance Policy Processes, 14th Stockholm Water Symposium, 16–20 August 2004, Stockholm

Gooch, G. D. and Huitema, D. J. (2008) 'Participation in water management: Theory and practice', in J. G. Timmerman, C. Pahl-Wostl and J. Möltgen (eds) *The Adaptiveness of IWRM – Analysing European IWRM Research*, IWA Publishing, London

Gooch, G. D. and Stålnacke, P. (eds) (2006) *Integrated Transboundary Water Management in Theory and Practice: Experiences from the New EU Eastern Border*, IWA Publishing, London

Groves, D. G. and Lempert, R. J. (2007) 'A new analytic method for finding policy relevant scenarios', *Global Environmental Change*, vol 17, pp73–85

Hope, R. A. (2006) 'Evaluating water policy scenarios against the priorities of the rural poor', *World Development*, vol 34, no 1, pp167–179

Kämäri, J., Alcamo, J., Bärlund, I., Duel, H., Farquharson, F., Flörke, M., Fry, M., Houghton-Carr, H., Kabat, P., Kaljonen, M., Kok, K., Meijer, K. S., Rekolainen, S., Sendzimir, J., Varjopuro, R. and Villars, N. (2008) 'Envisioning the future of water in Europe: The SCENES project', *E-WAter*, www.ewaonline.de/portale/ewa/ewa.nsf/home?readform&objectid=0AB6528C5177A8B7C12572B1004EF1C7, last accessed March 2010

Kok, K., Biggs, R. and Zurek, M. (2007) 'Methods for developing multiscale participatory scenarios: Insights from Southern Africa and Europe', *Ecology and Society*, vol 12, no 1, www.ecologyandsociety.org/vol12/iss1/art8/, accessed 17 June 2009

Liu, Y., Guo, H., Zhang, Z., Wang, L., Dai, Y. and Fan, Y. (2007) 'An optimization method based on scenario analysis for watershed management under uncertainty', *Environmental Management*, vol 39, pp678–690

Nakicenovic, N., Alcamo, J., Davis, G., de Vries, B., Fenhann, J., Gaffin, S., Gregory, K., Grubler, A., Jung, T. Y., Kram, T., La Rovere, E. L., Michaelis, L., Mori, S., Morita, T., Pepper, W., Pitcher, H. M., Price, L., Riahi, K., Roehrl, A., Rogner, H. H., Sankovski, A., Schlesinger, M., Shukla, P., Smith, S. J., Swart, R., van Rooijen, S., Victor, N. and Dadi, Z. (2000) *Special Report on Emissions Scenarios: A Special Report of Working Group III of the Intergovernmental Panel on Climate Change*, Cambridge University Press, Cambridge

Pahl-Wostl, C. (2002) 'Participative and stakeholder-based policy design: Evaluation and modelling processes', *Integrated Assessment*, vol 3, no 1, pp3–14

Pallottino, S., Sechi, G. and Zuddas, P. (2005) 'A DSS for water resources management under uncertainty by scenario analysis', *Environmental Modelling & Software*, vol 20, pp1031–1042, http://sorsa.unica.it/RO/Prof/pubblicazioni/pallo_sechi_zudd.pdf, last accessed March 2010

Peterson, G., Cumming, G. and Carpenter, S. (2003) 'Scenario planning: A tool for conservation in an uncertain world', *Conservation Biology*, vol 17, no 2, pp358–366

Seckler, D., Upali, A., Molden, D., de Silva, R. and Barker, R. (1998) *World Water Demand and Supply, 1990 to 2025: Scenarios and Issues*, Research Report 19, International Water Management Institute, Colombo, Sri Lanka

Shell (2003) *Exploring the Future. Scenarios: An Explorers Guide*, Shell International Limited, London

van Notten, P., Rotmans, J., van Asselt, M. and Rothman, D. (2003) 'An updated scenario typology', *Futures*, vol 35, issue 5, pp423–443

5

The Science–Policy–Stakeholder Interface in Water Pollution Assessment

Bruna Grizzetti, Antonio Lo Porto, Line J. Barkved, K. J. Joy, Suhas Paranjape, Johannes Deelstra, Fayçal Bouraoui and S. Manasi

Participatory watershed modelling: A tool for integration

During the last few decades, scientific research has increasingly developed modelling tools to handle problems related to water quantity and quality. Concerning water pollution, modelling approaches have been developed, ranging from simple statistical regressions to complex models, which describe the physical processes both at different levels of detail, and temporal and spatial resolution. Watershed models have been developed, first, to deepen the understanding of water cycle and pollution dynamics, as they provide a simplified conceptualization of the processes involved; and, second, as they have been increasingly used to evaluate the impacts of management options on water quality, mitigation measures and possible global changes. In addition, models have been employed for planning monitoring network and to spatially target plans for intervention. So far modelling has been exclusively used by the scientific and practitioner communities. However, there is a shift to a combined use of models by scientists, practitioners and stakeholders.

In a complex world, where natural resources and economic dynamics are firmly interconnected, the scientific community has recognized the need of more inter/transdisciplinary work in the process of knowledge production to understand the interactions between nature and society, and to support the sustainable use and management of natural resources (Kates et al, 2001). Effective scientific research needs to address questions relevant to stakeholders and decision-makers and to include their values and perspectives in the knowledge production process. This idea forms the basis of participatory watershed modelling. In contrast to traditional modelling, participatory watershed modelling includes both the public and decision-makers in the modelling process to support decisions involving complex environmental questions (Voinov and Gaddis, 2008). By including stakeholders' views and concerns, participatory modelling allows the incorporation of local knowledge and can legitimize the decision-making process (Korfmacher, 2001). Participatory modelling acts as a platform for dialogue and integration between scientists, stakeholders and policy-makers. In addition, the integration of stakeholders within the modelling process facilitates acceptance of the modelling outcomes. For these reasons its results are particularly pertinent to the objectives of IWRM.

In Europe the Water Framework Directive (WFD) (EC, 2000), together with the recent Marine Strategy Directive (EC, 2008) provide the integrated management structure for water policy, with the aim of protecting and enhancing the quality of water bodies and of promoting sustainable use of the water resource. The WFD requires member states to perform the analysis of pressures and impacts upon the water bodies at river basin level, including an economic analysis, and to set monitoring programmes to establish a coherent and comprehensive overview of the water status. In order to achieve the environmental objectives, member states have to develop and implement river basin management plans (RBMPs) for all river basin districts, including the programme of measures and economic strategies to recover the environmental cost of water services. For the implementation of the WFD, public participation is recommended at all stages of the planning process. Interested parties will be involved in information supply and consultation. In addition, member states will encourage their active involvement in the production, review and updating of the river basin management plans (CIS, 2003). Public participation allows a better collection of information for scientific assessment and policy-making, increases the possibilities of adapting measures to local conditions, and creates awareness and raises public acceptance and commitment for water management decisions (CIS, 2003).

As noted in Chapter 4, a way of involving public participation in water management is the development and use of scenarios. Scenarios are instruments for considering possible futures, which allow the identification of challenges and options in relation to different pathways of socio-economic development. They are used to assess the impacts of potential global and regional changes and to evaluate alternative management strategies and possible mitigation measures. Scenarios can either be formulated in the form of narrative stories, which are then translated in terms of quantitative changes, or consist of sets of measures related to environmental objectives. The effects on water quantity and quality associated with scenarios are quantified through the implementation of watershed models. Stakeholders' involvement in the development and application of watershed modelling is fundamental to the generation of meaningful scenarios and for promoting IWRM (Caille et al, 2007).

However, the development of scenarios and their evaluation through participatory watershed modelling involves many challenges related to the choice and application of the modelling approach; the availability of data and adequate monitoring network; the practical involvement of stakeholders; and the whole planning and implementation of the iterative process – all of which entails contrasting interests and communication barriers. The following section discusses these issues, providing examples and lessons learned from the practical experience gained in the case basins during the STRIVER project.

Watershed modelling with stakeholders in two case areas

A water pollution study was conducted according to the principles of participatory watershed modelling (Korfmacher, 2001) by twinning the Norwegian and the Indian study areas. More specifically, water quality in relation to diffuse nutrients pollution was studied in the Glomma, in the sub-basins Hunnselva and Lena, and in the Tungabhadra River basins. The same methodological approach was used throughout:

- implementing the same watershed modelling tool;
- involving the local stakeholders in the different stages of scenario development and analysis.

The study was conducted with the overarching objective of promoting mutual transfer of know-how (e.g. experiences, concepts and results) and technology (e.g. methodologies and models) in the twinned river basins.

The underpinning idea was to combine the potential of watershed modelling tools available in scientific research with the knowledge of local stakeholders, including water users and managers. In the project, three stakeholders meetings per basin were organized over a three-year period, contributing to different parts of the project. These meetings were the focal events during the project where researchers tried to enhance interactions with stakeholders, allowing their involvement at different stages of the modelling process (see Figure 5.1).

We consider here the challenges within the different phases of the modelling process by analysing the contribution of the stakeholders' involvement and referring to practical experiences during the project.

Phase I: Watershed modelling and data availability

The study of nutrient pollution at the river basin scale requires knowledge of the various sources and an understanding of their transport and transformation processes along the river basin. Watershed models are tools that enable the conceptual representation of the physical processes related to water quantity and quality at the river basin level, combining information on basin physical characteristics with pollution sources and processes dynamics. Through models, it is possible to estimate nutrient and sediment loads in water and to identify the contribution of different sources of pollution. Once validated, models can be applied to test the effects of alternative management options and are therefore used to select best management practices, or may be used to rank the costs

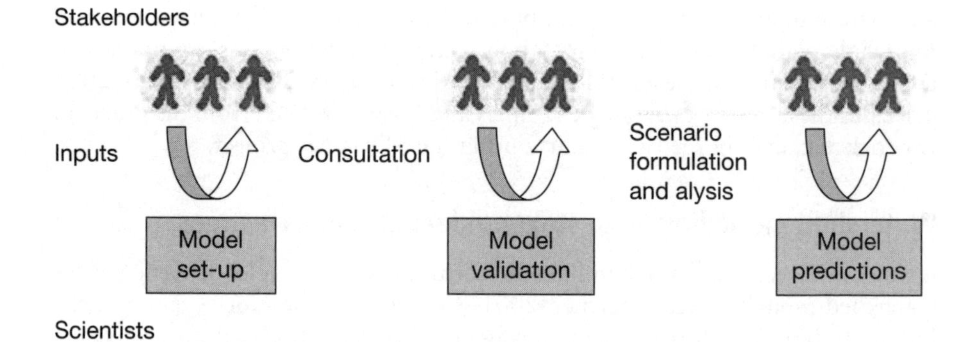

Figure 5.1 Scientist–stakeholder interaction during the modelling exercise

of different mitigation measures, providing relevant information for water managers and policy-makers.

A wide range of models exist, but all of them provide both advantages and limitations. In fact, modelling tools differ for the processes described regarding the data requirements, and the spatial and temporal resolution of inputs and outcomes (Silgram et al, 2008; Schoumans et al, 2009). In general, their complexity increases with the number of processes included and the resolution of predictions, as well as the timing of implementation and the expertise required.

The choice of a suitable model depends upon the objectives of the study (including spatial and time resolution), data availability, and the resources available in terms of budget, time and expertise. Selecting an appropriate model is not trivial since all of these factors need to be evaluated before the modelling exercise when, as is usually the case, not all the information is available; moreover, the choice involves a trade-off between model advantages and limitations. To support IWRM, it is paramount that the selected model can address the main identified pressures on the river basin and that the managers can trust the results of the modelling so that sound scientific advice and economically defensible decisions can be made based on the results. The quantification tools should aim to be accurate and responsive to changes in land use and land management. In addition, it is important to complete the model predictions by estimating their level of uncertainty in order to provide more reliable information for decision-making and regulatory formulation. Indeed, highlighting uncertainty during communication increases transparency and enhances the credibility of scientific support to decision-making. The whole spectrum of uncertainty has to be communicated, ranging from the uncertainty linked to the choice of model, to the model representation of the real world, to the data quality, and the risk that a decision-maker is willing to take to solve a particular problem.

As a general guideline, the choice of a suitable model should consider the availability of data. Many types of data are required to provide a picture of water pollution in a river basin and to apply watershed models. These data include the physical characteristics of the region of study (such as topography, river network, soils, aquifers, land cover, climate, lakes and reservoirs, etc.), the information on economic activities pollution sources (such as point discharges, agricultural areas and relative farming practices), and time series of measured water quality and quantity.

These data, when available, are often collected under different institutions or agencies within the same river basin, and they are produced using temporal and spatial scales that may not suit the modelling needs. In this context, the collection or the construction of a coherent set of data for modelling purposes provides the means for enhancing integration, both within the frame of IWRM and during the ensuing dialogue with stakeholders. In order to retrieve data, modellers may have to interface with different environmental agencies or research institutes, and they often struggle due to intra-institutional conflicts or barriers. At the administrative and political level, data gathering and distribution provide an overview on the degree of IWRM implementation in the river basin.

In the two case basins used in this chapter, the idea was to test two types of models, TEOTIL (Tjomsland et al, 2009) and SWAT (Neitsch et al, 2002), respectively. TEOTIL is a simple model that operates mainly on the basis of export coefficients for different land-use types. SWAT is a physically based model and relies on detailed spatial input data, such as soil, land-use and climate data. TEOTIL was chosen as a screening tool to identify areas with possible pollution problems, while SWAT was selected to perform scenarios analysis. TEOTIL requires information on nutrient sources on the sub-basin level and estimates nutrient source apportionment at sub-basins outlets on an annual basis. SWAT is more demanding in terms of time and data input as it requires part of the information on daily time steps. However, in addition to nutrient source apportionment and spatial mapping of water pollution, SWAT provides estimates of daily water flow and nutrient concentration at the sub-basin level and is able to simulate the impact of climate or land-use changes on water quality and quantity. During this project, the overall modelling performance was found to be reasonable, although shortages and gaps in the required data were identified during the modelling process, leading to assumptions or data constructions. More details on the modelling results for the Glomma and Tungabhadra river basins are reported in Barkved et al (2008) and Lo Porto et al (2008), respectively.

The practical implementation of watershed modelling in the two case river basins used in this chapter highlighted the challenge of quantifying water pollution pressures when coping with data scarcity. The work benefited from the involvement of stakeholders in the modelling process. In fact, stakeholders provided a significant input in identifying water pollution pressures and prioritizing them based on their perception and local knowledge. In addition, stakeholders contributed substantially to the gathering of data and information collection.

Although the choice of modelling tools was already determined at the beginning of the project, the stakeholders were provided with the opportunity to discuss what type of water quality pressures and processes they were interested in. Building upon the potential of the two modelling tools, scientists and stakeholders interacted to explore the possibilities of investigating the main sources of river basin water pollution and improving targeted project modelling activity. From discussions held with stakeholders, it was evident from the very beginning that both the models proposed had the potential to answer relevant questions for the local water managers and water users. However, the data needed by the models were not fully available in the regions of study, thus limiting the models' application and increasing the uncertainty of their outcomes. Consulting the stakeholders would therefore have been beneficial before the choice of the modelling tools was decided. It would have raised more awareness on the issue of data scarcity in the regions of study. However, by selecting technical tools from the start, the resulting stakeholder dialogue benefited from being able to show practical solutions and structure the discussion, resulting in a proactive approach to start the interaction.

In this project one of the main challenges during the first phase of the modelling exercise was the availability of monitoring data for water quantity and quality. Some other limitations were encountered regarding the availability of data for model parameterization, such as weather time series, soil characteristics, land cover and farming practices. At this early stage, the discussion with stakeholders helped in collecting relevant information for modelling, such as local agricultural practices or rate of connection to sewerage networks. In addition, in some cases it was possible to utilize several global coverage data sources, such as in the Tungabhadra River Basin. In some other cases, modellers had to infer missing parameters from proxy data and modelling equations, which was the case for gaining, for example, soil information in the Hunnselva and Lena river basins. However, data scarcity might lead to larger uncertainty in model predictions. Scientists therefore explained the limitations imposed by the lack of data, and this was understood and considered by stakeholders in the following discussions on management scenarios and associated uncertainty. This experience raised the stakeholders' awareness regarding the importance of reliable monitoring activities to characterize the status of water resources; evaluating the effect of mitigation measures; revealing the occurrence of new threats; and supporting modelling scenario studies.

Examples are given below of how stakeholders can contribute to the various modelling stages and include the identification and control of water pollution for the Indian case study, and data collation in the Norwegian case study.

Stakeholder contribution in pollution identification and control: The Tungabhadra example (India)

Although periodic pollution monitoring is the responsibility of the state in India, the experience in the Tungabhadra sub-basin shows that civil society initiatives have also had a significant role to play in acting as a watchdog over the state's efforts. The primary agencies responsible for monitoring control and prevention of water pollution at the state level are the respective state pollution control boards. In the case of Tungabhadra River, two state pollution control boards are involved – namely, the Karnataka State Pollution Control Board (KSPCB) and the Andhra Pradesh State Pollution Control Board. The research team had the opportunity to interact with the KSPCB extensively and many of their officers have actively participated in the stakeholder meetings.

The reactions during the stakeholder meetings to the performance of the KSPCB were mixed. Some of the stakeholders clearly felt that because of the various measures taken by the KSPCB in terms of active monitoring of the hotspots, water quality had improved over the years. Of course, the KSPCB has taken proactive steps only because of pressure from civil society organizations. There was also another opinion voiced: that more could still be done by the KSPCB.

During the second stakeholder meeting at Davanagere, a representative of the Environmental Office of Shimoga District spoke on the efforts being made by the KSPCB for improving water quality. For example, efforts were made to reduce the effluents discharge from the steel and paper mill industries. In addition, under the National River Conservation Plan, efforts are being made to erect sewage treatment plants in the towns of Bhadravati and Shimoga, and to construct both community and individual toilets and crematoriums for the poor people living on the banks of the river. These efforts, though fragmented, have been necessitated mainly by the constant agitation of, and protests by, the people within the basin, led by local organizations. Besides fighting against pollution, local organizations also took the initiative of undertaking scientific studies and submitted the reports to KSPCB. A case was registered in the Karnataka High Court and, based on the conditions of fisheries and occupational health, the court asked the industries to clean up the river. A local watchdog committee was also formed to monitor the pollution control measures of KSPCB. Thus, one can say that the basin has a vibrant and informed civil society movement against pollution.

From the project stakeholders meetings and from many of the reports of the Central Pollution Control Board, it emerged that the water quality in the Tungabhadra River Basin is affected by four major activities in the basin:

1 industry;
2 mining;
3 urban and rural settlements and their waste disposal systems;
4 runoff from agricultural fields.

Water quality along the river is being monitored at various places under different programmes. However, only a few parameters are consistently being monitored over sufficiently long periods. Trends show that upstream stretches are relatively less polluted than the downstream stretches. The data also show that average water quality deteriorated dramatically during the late 1990s and has subsequently improved. However, the data range has also simultaneously increased, implying that there are periods when the pollution levels are quite high. Downstream pollution levels have also not improved much over their late 1990 values. Fish kills and deterioration in fish catch have also been reported, although there are indications that the actual adverse impact may be a compound phenomenon that is affected not only by water quality *per se*, but also by changes in fishing technology and practices.

In Tungabhadra, much of the improvement may be related to the rise of a vibrant citizens' initiative against pollution during the 1990s, leading to improved treatment of the effluents arising from the chemical industry. However, there is a need both for closer monitoring and improving stakeholder participation in the process. There is now greater awareness of water quality issues amongst the citizenry in the area, although it is sometimes only narrowly focused on the two major activities of industry and mining. This awareness needs to be extended to two other factors – namely, agricultural diffuse pollution and urban wastewaters, as well as the associated changes required. There is a need for continued monitoring, as well as vigilance, on the part of civil society groups and concerned stakeholders. Closer participative monitoring feeding into a participative basin management is of great importance for improving and maintaining water quality in the basin.

Stakeholders' contribution in data gathering: The Hunnselva/Lena example (Norway)

The European Union Water Framework Directive (WFD) is currently under implementation in Norway with a six-year delay compared to the rest of Europe. Some selected basins, such as Hunnselva, are following the 'normal' European time frame, meaning that the WFD has to be implemented by 2015. Both the WFD and the licensing systems in Norway's national legislation have developed frameworks for stakeholder participation. Project scientists from the onset of the project established a close collaboration with the Hunnselva River Basin Working Group for the implementation of the WFD, including water managers in central, regional and local governments. The contacts were both formal and informal in character. Using the Hunnselva and the neighbouring Lena Basin as a case for the pollution study enabled project participants to connect to a 'real case' and interact with an ongoing process and study on the WFD implementation process from an IWRM point.

Through this collaboration the project participants had the chance to meet and discuss water pollution aspects over a longer period of time. The scientist–policy–stakeholder interaction in the basin progressed successfully mainly due to several encounters between managers and scientists during the project period, enabling the progress of a gradual mutual understanding of the scientific and management aspects. An important lesson learned from this is that sufficient time (and resources) should be allowed for this process to occur and for the stakeholders (the parties involved) to digest the information. In fact, building trust, mutual understanding and a 'good space for interaction and communication' takes time and effort. However, care should also be taken that time should not be spent on endless debates, but rather on establishing a clear time frame within which such a process occurs and to visualize the outcomes in a timely manner.

In the application of the modelling tools in Hunnselva and Lena (sub-basins of Glomma), the lack and inconsistency of data proved to be a constraint. This applies to environmental as well as economic and social data. The main limitations concerning data availability were that data were not always available at the required scale and resolution needed for model calibration and validation. At the same time, data were rarely owned by the same institution. In Hunnselva and Lena, there is a lack of coordination between water quantity (hydrological) and quality monitoring. This emphasized the ongoing national discussion in Norway on the need for harmonized water quantity and quality monitoring to support an integrated river basin management approach. For the Hunnselva case, in particular, the establishment of a water flow gauge in the lower parts of the catchment is a minimum requirement (also pointed out by the local WFD group). During the working process, data availability and additional collection were discussed with the stakeholder groups. A field trip was organized in the basins to study basin characteristics, problem areas and agricultural practices and to discuss possible scenarios for the modelling. The field trip formed an important input to the final formulation of scenarios for the modelling. As a follow-up, based on the information gathered during the field trip, a questionnaire was used to obtain and update the necessary information from the relevant stakeholders and to identify the scenarios to be modelled.

The STRIVER exercise showed that in Norway, given the fragmented institutional arrangements for water management, data are collected by different organizations for different reasons, in different ways and with variable approaches to data storage. When scientists in such a context try using existing data sets, it is still not uncommon to find that insufficient spatial and temporal resolution make the analysis difficult. Furthermore, our experiences emphasized the importance of involving stakeholders in the data processing of the model exercises, incorporating local knowledge and understanding the natural system and ensuring local participation in decision-making. The interactions with the stakeholders placed focus on data limitations from a management perspective and provided scientists with local knowledge of the conditions, practices and management of the basins. The modelling exercise may be said to have provided a framework for interpreting data, integrating information and identifying gaps in data or current knowledge. The interactive process of data acquisition created awareness among the participants of the 'conditions' in the basin. In addition, scientists could draw on and resort to (expert) local knowledge in the process.

Phase II: Scenario development and analysis

Scenarios are instruments to evaluate possible future directions for development or policy implementation (see Chapter 4). They consist of a description of possible policy measures or global and regional changes. Scenarios can be formulated in the form of narrative stories, which are then translated in terms of quantitative changes, or consist of sets of measures related to environmental objectives. The effects of scenarios on water quality and quantity can be assessed by the implementation of watershed models. Different social actions and policy measures are translated into the relevant changes in model parameters, allowing the exploration of the impact of different policy options and their combinations on river flow quantity and quality over time and space.

Scenarios for water quality need to include the knowledge, mind-frames and interests of different stakeholders, and should build upon an understanding of the present-day situation. The trialogue between scientists (modellers), policy-makers and the other stakeholders involved is of fundamental importance to the development of meaningful and relevant scenarios that are likely to be used by practitioners. The role and mechanisms of stakeholder involvement in scenarios development are thoroughly discussed in Chapters 3 and 4. The participation of stakeholders in the modelling process provides a substantial input for scenario development. The involvement of stakeholders contributed to:

- the development of realistic scenarios that correspond to the threats and interests of the different parts represented by the stakeholder group;
- analyses of the results of scenarios predictions, providing feedback to the scientists and the decision-makers;
- the development of trust between scientists and stakeholders;
- an increase of awareness and acceptance of consequences of social behaviours, economic measures or policy implementation.

The scenario modelling in the two case study areas showed how the interest of stakeholders was driven by water pollution issues and data constraints specific to each river basin, indicating the relevance of including local knowledge.

Modelling scenarios with stakeholders: The Tungabhadra case (India)

In the Tungabhadra Basin in India, an intensive stakeholder interaction preceded the scenario building phase. The first stakeholder workshop was held at Hospet, near the Tungabhadra Dam in Karnataka State in 2007. There were two follow-up workshops organized and held at Davangere in Karnataka State and Kurnool in Andhra Pradesh. This was ensued by the second stakeholder workshop organized at Bangalore in 2008. The main inputs to the scenarios were provided during this second meeting. Farmers, state officials, policy-makers, scientists, civil society organizations and innovative farmers participated in the workshop. Research findings and requirements were presented to the stakeholders. The concept of scenarios was discussed, including how different options could be included. At the workshop, it was felt that the research findings provided a picture of the actual situation of the water quality in the river basin and 'business as usual' scenarios could be built quite easily from observed trends. In contrast, it emerged that the development of a realistic IWRM scenario that could be perceived as achievable and acceptable by the stakeholders needed to be discussed more intensively.

Three break-out groups were formed. Given the baseline situation as described by the scientists in their presentations on the first half day, the three groups were asked to deliberate upon the following five key questions:

1 What are the different options to reduce industrial, environmental and agricultural pollution – for example, polluting inputs such as chemical pesticides and fertilizers?
2 How can water used for agriculture be reduced, especially regarding the three major crops of paddy, areca nut and sugar cane?
3 How should urban and rural areas deal with the provision of sanitation services and sewage disposal?
4 How should allocation to different water uses be prioritized?
5 What kinds of institutions should be created to accommodate the different stakeholders and the different scales of water management – micro-watershed, watershed, sub-basin, basin, village, town, city, state?

The participants were asked to consider their answers carefully, and to be realistic in their assessments of how and how much the different measures could influence results. The result was a set of promising IWRM options and the relative policy support required. Some of these options were then tested as scenarios through watershed modelling. Scenario topics particularly addressed related to water demand for different uses; cropping pattern and agronomic practices; treatment of effluents and urban sewage and sanitation practices in rural areas; trends in urbanization; and impacts of climate change. The qualitative scenarios (from a ten-year perspective) were then translated into the following quantitative scenarios:

- *Climate change*: climate variations according to the Regional Climate Model Hadrm3 and emission scenarios A2 and B2 (worst and best future global CO_2 emissions scenarios, respectively).
- *Sewage treatment*: 80 per cent of households adopt septic tank sanitation in rural areas and 50 per cent of urban sewage undergoes treatment.
- *Irrigation system*: sprinkler system is implemented in 80 per cent of orchard crops.
- *Crop system*: 40 per cent of the irrigated area upstream of the Tungabhadra Dam and 10 per cent downstream shifts to the System of Rice Intensification (SRI).

The modelling results of the climate change scenario indicated that under the emissions of A2 global scenario (worst climate change), the sediments load in the river system will double compared to actual conditions, inducing a sediment inflow into the Tungabhadra Dam three times higher than the present one, while the sediment outflow will remain quite stable. This will lead to a substantial increase in siltation rate and reduce the dam's life span. Nutrient balance in the river will not change much compared to baseline figures. The sewage treatment scenario showed that a 50 per cent reduction from sewage discharge can be expected in both nitrogen and phosphorus contribution to the overall river pollution as a result of improved sanitation in rural areas and urban wastewater treatment. Concerning the crop system, changes in the spatial pattern of the System of Rice Intensification will result in a 6 per cent saving of the river basin water yield, accompanied by an overall increase of rice yield of about 20 per cent. The outcomes from the scenario on irrigation changes were considered to be affected by too high an uncertainty, since data on cropping pattern and practices were not available at the appropriate resolution.

The modelling results were well received by the stakeholders. The results also implied that the course of action in the Tungabhadra Sub-Basin (TBSB) had to focus on:

- improving the data collection and monitoring process within the basin;
- erosion control as a high priority area;
- rapid improvement of sewage disposal and treatment facilities;
- diffusion of new techniques (e.g. in rice cropping: SRI) that play a relevant role in saving water resources.

These matched well with stakeholder suggestions and reinforced them. The finding on the rate of siltation was especially striking. Many stakeholders felt that the greatest threat in this respect was the mining activity, especially the illegal mining taking place in and around the Tungabhadra Dam. The stakeholder's forum that has resolved to continue its activity in the post-project phase intends to take up many of these issues, including that of mining, in its subsequent activity.

Modelling scenarios with stakeholders: The Hunnselva and Lena case (Norway)

In the Glomma River Basin, three stakeholder meetings were held during the project's duration. A number of one-day field trips were also organized by the project researchers to meet the local stakeholders *in situ*, such as farmers and fishermen. During the stakeholders' meetings, issues related to water pollution sources and water quantity distribution among users and sectors were addressed. In the case of the Hunnselva and Lena catchments, stakeholders' interests were mainly focused on the impact of measures relating to agricultural practices, as required by the new environmental policies. Together with local stakeholders, potential scenarios relating to nutrient losses from agriculture were defined. The scenarios consisted of two possible mitigation measures and also an exploratory 'what if' type of scenario, as follows:

- *Reduced tillage*: the autumn ploughing is substituted with a light harrowing.
- *Optimal fertilization*: current crop fertilizer applications are replaced by optimal fertilizer applications, set according to the actual crop needs.
- *Crop change*: the growing of cash crops (vegetables) is extended by 5 per cent.

The modelling results indicated that the introduction of reduced tillage would decrease both soil erosion and phosphorus pollution at the catchment's outlet. Similarly, the application of more balanced fertilization to all crops will lead to a decrease in nutrient losses, especially nitrogen. Finally, modelling showed that a land-use shift of 5 per cent from barley to vegetables, while being more economically advantageous on the Norwegian market, would increase nitrogen losses, whereas the impact upon phosphorus would not be immediately evident.

These types of outcomes were considered to be relevant by local stakeholders, who are facing the introduction of a new water policy and are therefore concerned about the link between agricultural revenues and environmental measures. Stakeholders realized the potential of scenario modelling to address their concerns, as well as the dependence of modelling on data quality and availability. The interaction with stakeholders resulted in defining realistic scenarios, providing information on the impact of the measures they were asked to implement by the coming legislation, or changes they were willing to make as a consequence of market-driven conditions. Therefore, the scenarios were very practical and responsive to stakeholders' needs, who in this case were mainly farmers and water managers. The results have subsequently been used as supporting knowledge in planning mitigation measures in the basin.

In addition, stakeholders' perception of the main polluting sources affecting water quality in the river basin provided an important input into the modelling set-up and helped in seeking the relevant information. The interaction with modellers also increased the stakeholders' awareness of the importance of data availability and monitoring plans to check water quality conditions in the long term.

Lessons learned in the STRIVER experience

Stakeholders were involved in all phases of the modelling process and scenario-building in the project. Their participation was beneficial throughout the whole study. In the initial phase, the interaction with stakeholders was fundamental to help set water pollution priorities, and to understand water quantity distribution among users and sectors. At this primary stage, the stakeholders' knowledge of local conditions, and their perception of the main water polluting sources in the river basin, provided important input to the modelling set-up and helped to complete or integrate the technical data required. During the second phase, the scientist–stakeholder interaction proved beneficial to the process of validating the model representation of river basin conditions, particularly in the formulation of alternative management scenarios relevant to local people. In addition, scenarios were presented and discussed with the stakeholders, increasing their awareness of measures and policy implementation, as well as uptake of the scientific outcomes (see Figure 5.1).

The scientist–stakeholder interaction was conducted as an iterative process throughout the various stages of the modelling exercise, and within the project's budget and time frame. While simple in principle, the interaction between scientists and stakeholders in the modelling process encountered some practical difficulties. For example, the use of group-specific terminology created communication problems between the different actors involved. Moreover, scientists, water managers, policy-makers and different groups of water users hold different priorities and approaches, which further challenges their integration together as a working group. Other practical difficulties were linked to the frequency and scheduling of the stakeholder meetings as they did not always fit in with the working time required for the modelling implementation, and the physical distance between modellers and stakeholders, who were based in different countries, or even different continents. The issue of data availability appears to be crucial for supporting watershed modelling and scenario development. In the two case studies, the data for watershed modelling were not always available at the required scale and resolution, and were rarely owned by the same institution. In addition, water managers were not always aware whether the current monitoring network was adequate to answer their requirements. Improving the dialogue between water managers and the research community would help to recognize research needs, make available the existing knowledge, identify knowledge gaps and plan data collection and monitoring programmes.

In general, then, the involvement of stakeholders in all phases of the watershed modelling process and in the development of scenarios resulted in a positive experience, which improved the quality of the analysis and results, and provided valuable lessons for similar processes (Grizzetti et al, 2008). In particular, experiences in the two case basins have shown that involvement of stakeholders in the modelling process significantly contributed to:

- prioritization of water quality and quantity problems, targeting the modelling objectives;
- inclusion of local knowledge in the process, ensuring more reliable results of the modelling activity;
- developing scenarios of real interest for the parts represented by the stakeholders;
- analysis of the results of scenario predictions, providing feedback to scientists and decisions-makers;
- building trust between scientists and stakeholders, and improving transparency;
- increasing stakeholders' acceptance of measures and policy implementation.

Watershed modelling with stakeholder participation can respond effectively to the water quantity and quality issues of IWRM, acting as the platform for a trialogue between science, policy and stakeholders and supporting the development of integrated solutions by including local knowledge.

References

Barkved, L. J., Deelstra, J., Grizzetti, B. and Bouraoui, F. (2008) *Modelling Nutrients in the Glomma River Basin: Scenarios and Management Recommendations*, STRIVER Policy Brief No 11, http://kvina.niva.no/striver/Portals/0/documents/STRIVER_PB11.pdf, last accessed March 2010

Caille, F., Riera J. L., Rodriguez-Labajos, B., Middelkoop, H. and Rosell-Mele, A. (2007) 'Participatory scenario development for integrated assessment of nutrient flows in Catalan river catchment', *Hydrology and Earth System Sciences Discussions*, vol 4, pp1265–1299

CIS (2003) *Common Implementation Strategy for the Water Framework Directive (2000/60/EC): Guidance Document No 8, Public Participation in Relation to the Water Framework Directive*, Produced by Working Group 2.9 – Public Participation, Office for Official Publications of the European Communities, Luxembourg

EC (2000) *EU Water Framework Directive, Directive 2000/60/EC of the European Parliament and the Council of 23 October 2000 Establishing a Framework for Community Action in the Field of Water Policy*, Official Journal of the European Communities (22.12.2000) L 327/1)

EC (2008) *Directive 2008/56/EC of the European Parliament and of the Council of 17 June 2008 Establishing a Framework for Community Action in the Field of Marine Environmental Policy (Marine Strategy Framework Directive)*, Official Journal of the European Union, 25.6.2008, L 164/19–164/40

Grizzetti, B., Lo Porto, A., Barkved, L. J., Bouraoui, F., Deelstra, J. and Joy, K. J. (2008) *Modelling Water Pollution with Stakeholders' Involvement – The Twinned Experience of Glomma (Norway) and Tungabhadra (India) River Basins*, STRIVER Policy Brief No 10, http://kvina.niva.no/striver/Portals/0/documents/STRIVER_PB10.pdf, last accessed March 2010

Kates, R. W., Clark, W. C., Corell, R., Hall, J. M., Jaeger, C. C., Lowe, I., McCarthy, J. J., Schellnhuber, H. J., Bolin, B., Dickson, N. M., Faucheux, S., Gallopin, G. C., Grübler, A., Huntley, B., Jäger, J., Jodha, N. S., Kasperson, R. E., Mabogunje, A., Matson, P., Mooney, H., Moore III, B., O'Riordan, T. and Svedin, U. (2001) 'Sustainability science', *Science*, vol 292, no 5517, pp641–642

Korfmacher, K. S. (2001) 'The politics of participation in watershed modelling', *Environmental Management*, vol 27, no 2, pp161–176

Jouvenel, H. (2000) 'A brief methodological guide to scenario building', *Technological Forecasting and Social Change*, vol 65, no 37, pp37–48

Lo Porto, A., Barkved, L. J. and Gosain, K. A. (2008) *Modelling Water and Nutrients Balance in Tungabhadra River Basin: Scenario Analysis and Management Recommendations*, STRIVER Policy Brief No 7, http://kvina.niva.no/striver/Portals/0/documents/STRIVER_PB7.pdf, last accessed March 2010

Neitsch, S. L., Arnold, J .G., Kiniry, J. R., Williams, J. R. and King, K .W. (2002) *Soil Water Assessment Tool Theoretical Documentation*, Grassland, Soil and Water

Research Laboratory, Agricultural Research Service, Blackland Research Center, Texas Agricultural Experimental Station, Temple, Texas

Schoumans, O. F. M., Silgram, P., Groenendijk, F., Bouraoui, H. E., Andersen, B., Kronvang, H., Behrendt, B., Arheimer, H., Johnsson, Y., Panagopoulos, M., Mimikou, A., Lo Porto, H., Reisser, G., Le Gall, A., Barr, S. and Anthony, G. (2009) 'Description of nine nutrient loss models: Capabilities and suitability based on their characteristics', *Journal of Environmental Monitoring*, vol 11, pp506–514

Silgram, M., Anthony, S. G., Fawcett, L. and Stromqvist, J. (2008) 'Evaluating catchment-scale models for diffuse pollution policy support: Some results from the EUROHARP project', *Environmental Science and Policy*, vol 11, no 2, pp153–162

STRIVER (2009) *Strategy and Methodology for Improved IWRM: An Integrated Interdisciplinary Assessment in Four Twinning River Basins*, Founded by the European Commission under the Sixth Framework Programme (SUSTDEV-2005-3.II.3.6: Twinning European/Third countries river basins), www.striver.no, last accessed March 2010

Tjomsland, T., Selvik, J. R. and Brænden, R. (2009) *Teotil Model for Calculation of Source Dependent Loads in River Basins*, NIVA, Oslo

Voinov, A. and Gaddis, E. J. B. (2008) 'Lessons for successful participatory watershed modelling: A perspective from modelling practitioners', *Ecological Modelling*, vol 216, pp197–207

6

The Science–Policy–Stakeholder Interface (SPSI) in Land- and Water-Use Interactions

S. Manasi, K. J. Joy, Suhas Paranjape, Udaya Sekhar Nagothu, Dale Campbell, N. Latha, Maria Manuela Portela, António Betamio de Almeida, Marta Machado, K. V. Raju and Santiago Beguería Portugues

Understanding land- and water-use interaction

Land use has significant impact upon water resources in terms of water quantity and quality and is, in part, determined by environmental factors such as soil characteristics, climate, topography and vegetation. Land and water characteristics are connected; yet while land characteristics tend to be relatively fixed in time, and spatially extensive, water characteristics vary much more in time but tend to be spatially concentrated, such as at points of measurement of stream flow or water quality. Land and water resources management needs to be integrated since the type of land use and management has implications for water and vice versa, and eventually on production, efficiency and livelihoods in a river basin. The relationship between land use and water quantity and quality is mutually dependent; land-use changes not only have a major impact upon water resources, but also have great potential for modifying the hydrological cycle within the river basin. However, most hydrological analyses do not emphasize the integration of water and land use, although there is much experimental evidence of the importance of land use on water resources generation (Bosch and Hewlett, 1982; Sahin and Hall, 1996; Beguería et al, 2006).

Starting with this point of departure, this chapter provides a comparative analysis of land and water management interactions and of their impacts in two river basins: the Tungabhadra Sub-Basin (TBSB) in India and the Tagus Basin in Spain and Portugal. The focus of the chapter is on the problems of land use and water management faced by practitioners and scientists in the context of integrating knowledge and experiences from scientific, managerial (policy) and stakeholders' (the science–policy–stakeholder interface, or SPSI) perspectives. An attempt is made to identify the main actors and to analyse the interaction or lack of interaction between them in the context of the case basin problems, as well as the communication of the problems between the actors. We also address the information available at various spatial levels and the mechanisms of information exchange between actors, keeping in mind data availability and the type of data required. Thus, the analysis provides insights into the kind of information/interventions that are in place or missing and the need for improved interactions between actors.

Land-use changes

In the TBSB, farmland and grazing land dominates the landscape, except for the dense forest patches on the headwater areas in the Western Ghats region, to the south-east of the region. According to municipal land-use statistics, forest cover has increased in the basin (Beguería et al, 2008). In the case of the Tagus, the increase of the urban areas has been the most significant land use change in the Spanish part of the basin, whereas in Portugal the increase and densification of the natural vegetation cover has been predominant (Beguería et al, 2008). However, in both Spain and Portugal, important urban areas have developed around the two main cities located in the Tagus Basin: Madrid and Lisbon, the capitals of Spain and Portugal. A general process of abandonment of the marginal lands for agriculture and pastures has also been observed in the headwaters of the Tagus River in Spain, and some new irrigated areas have been developed in the valleys of major tributaries. The time series of satellite-derived Normalized Difference Vegetation Index (NDVI) has also revealed a decrease of the vegetation around the Madrid area, contrasting with increased vegetation activity in the headwaters at the Iberian range and the Portuguese areas (Beguería et al, 2008).

This increase of the natural vegetation surface area and activity is known to cause an increase of the use of 'green water', or water consumption, by the ecosystem to maintain the ecological status. As a consequence, there is a reduction of the quantity of 'blue water', or water in rivers and lakes and in the underground aquifers which can be used by humans. This is compensated for by the positive effect of forests in regulating the water cycle. On the other hand, reduction of the forest cover and urbanization is known to increase (in some cases dramatically) the 'runoff coefficient', leading to increased soil erosion and a higher severity of floods. In the TBSB, assessment of the effects of land-use change at the basin level was difficult due to the lack of appropriate data in some cases.

In the TBSB, in addition to an increase of the vegetation activity in the headwaters, a negative trend in the annual rainfall was observed. Both conditions are expected to have had a negative effect on runoff production and river discharge. On the other hand, the analysis of reservoir storage time series revealed no impact of either land-use change or climate variability, showing a stationary time series only subject to natural year-to-year oscillation. These results suggest that in the TBSB, despite significant land-use change in the headwaters, which added to the effects of a reduction of precipitation in the last decades, the Tungabhadra reservoir system had a large resilience and no effect was apparent in the series of water storage. In the Tagus Basin, on the contrary, the regulation capacity at the headwaters is very high (more than 100 per cent of the annual water contribution). As a consequence, the system was much more sensitive to changes in the hydrological cycle. The analysis showed only a marginal effect of land-use change on the river discharge, although the time series of reservoir storage showed negative trends. In the Portuguese Tagus River Sub-Basin and for the time being most of the precipitation time series do not exhibit significant trends.

Main actors in the basins

As new demands for water emerge, pressure to reallocate water increases, leading to conflicts across and within sectors affecting economic and environmental prerogatives. While there is competition for access to water, different uses of water are not always mutually exclusive. In the Tagus and in the Tungabhadra river basins, different sectors/actors compete for water resources – namely, agriculture, industry, urban settlements, mining and hydropower production, and general socio-economic development. These are the main actors, although the role played by each sector varies according to the local conditions. In the TBSB, management is based on administrative and not hydrological boundaries, resulting in various allocations, distribution and usage problems within and across sectors. In the Tagus River Basin, water management is mainly based on the natural river basin boundaries, although the main user sectors are structured at different spatial levels: local, regional, national and (in terms of transboundary rivers, such as the Tagus River) international levels.

Among the various users of the TBSB, the main active actors are the farmer groups and industrialists, while the urban towns and fishing communities are the dispersed beneficiaries whose activities are not as collective in terms of access and implications. In the Tagus, in both Spain and in Portugal, water abstraction for irrigation, urban supplies and industrial discharges affect mainly water quantity, while in the Tungabhadra, both water quantity and quality are threatened by competing sectors. Industries and corporations impair the sustainable use of water resource through constant subtractions from surface and groundwater bodies and local discharges of polluted effluents. Hence, although existing in different contexts, the two river basins face, in some cases, similar pressures and comparable impacts.

Problems in land and water use

Land cover and land-use changes have had an impact upon the quantity and quality of both surface and groundwater resources by reducing environmental

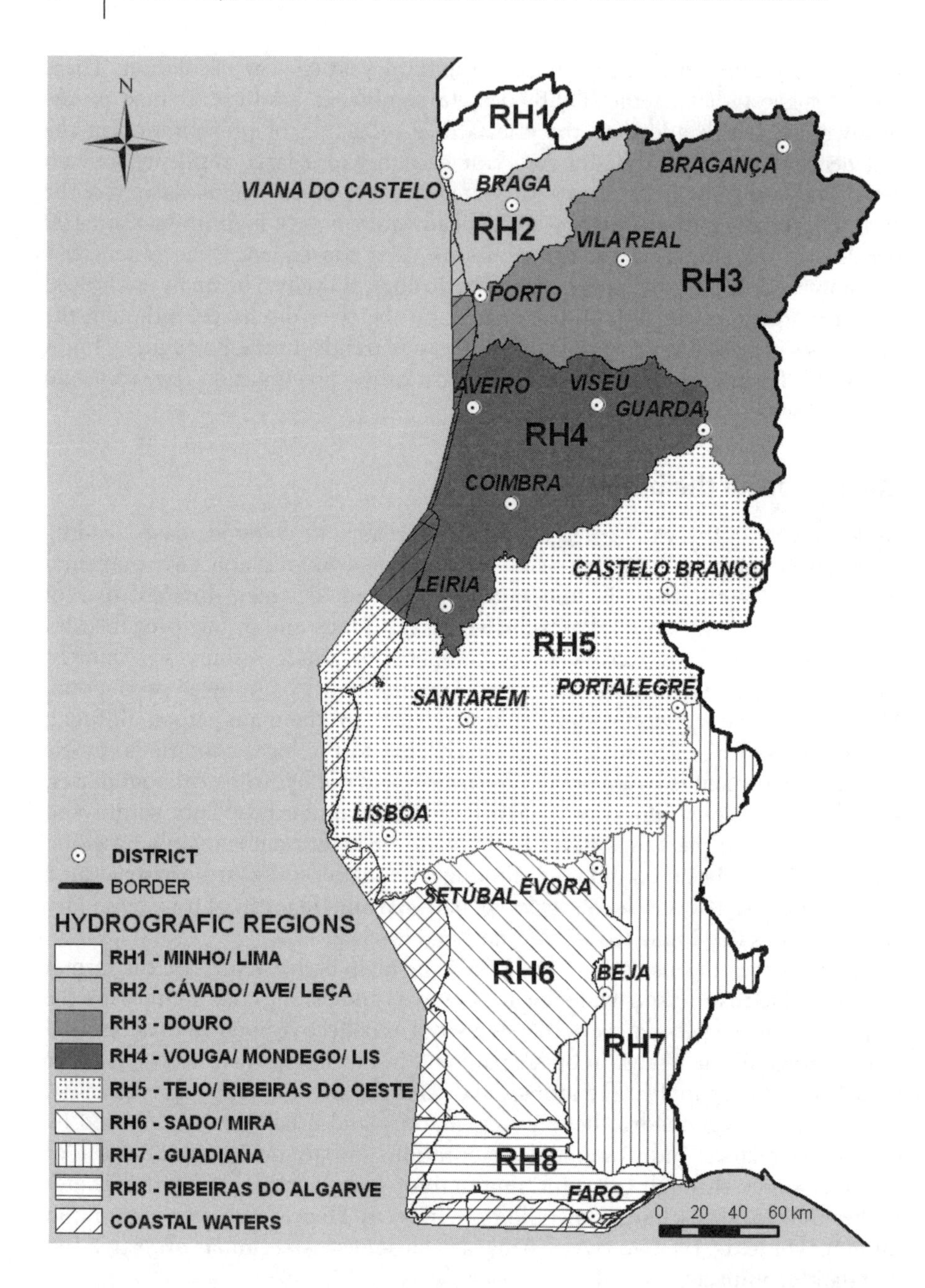

Figure 6.1 General location of the ten Portuguese hydrographic region administrations (HRAs)

Note: Besides the shoreline and the border, the limits of the HRAs are essentially coincident with the limits of the main Portuguese river basins.

Source: INAG, Instituto da Água, 2005. Relatório síntese sobre a caracterização das Regiões Hidrográficas prevista na Directiva Quadro da Água. Ministério do Ambiente, do Ordenamento do Território e do Desenvolvimento Regional, Lisboa, Portugal. p175

flows and river discharges, by increasing pollution, and by changing the properties of the land surface itself. They have also altered the hydrology and chemistry of the river basins. Water demands in the two basins are influenced by the natural variability of hydrological cycles, expansion of irrigated agriculture, and changing technology and governance regimes. Forest degradation resulting from felling trees and mining activities has led to a decline in biodiversity and habitat, and has contributed to flash floods and the pollution of rivers and land surrounding the watercourses. Frequent fires in the Tagus Basin have caused considerable destruction, while government agencies have undertaken partially successful reforestation works to attempt to mitigate such damage.

The agricultural sector consumes a major share of water in the TBSB and supports 80 per cent of the population in the region. Although the area is best suited to semi-arid crops, paddy and sugarcane are the major crops grown. Dependency on groundwater for irrigation has increased over time. In the TBSB, irrigated agriculture has replaced more traditional farming methods in some places, and now both rain-fed and irrigated farming methods co-exist. Irrigated and rain-fed farming are managed in isolation and there are few signs of integration within these domains. However, there are also a number of small-scale initiatives that serve as a starting point for integrated approaches, such as watershed programmes and land-use planning for integration. There is, however, still a need for alignment and widespread participation in these initiatives. In the Tagus Basin, there has been a decrease in dry-land farming and an increase in irrigated agriculture which has occurred independently of the increase in, and densification of, natural vegetation cover. Most of the basin water resources are also used for agriculture needs. As noted above, in the TBSB there are areas of highly intensively irrigated agriculture; however, the System of Rice Intensification (SRI) and aerobic rice cultivation are also practised in some parts of the basin, which produces similar yields while reducing water use. Similarly, in the Tagus Basin, many Spanish farms are excessively irrigated, which results in pollution from nitrates and other effluents.

Rapid urbanization and population pressures contribute to conflicts in the TBSB, together with pollution from industry, agriculture and urban sewage. Recently, there has been an increase in sensitivity to the issue of pollution, and regular monitoring by the Pollution Control Board is now conducted. One of the major sources of conflicts in the basin has been between the industries and the farmers, due to both industrial pollution and also to the use of water by the industries (inter-sectoral conflicts). Conflicts due to construction of dams in the basin are no longer so much of an issue, as the dams on the river are older and no new dam construction is planned. However, conflicts between downstream and upstream farmers in the service area (command area) of the existing dams do arise quite often, as many downstream farmers do not get sufficient access to water.

Concerning flood control, the Bhadra and Tungabhadra dams are now completed but are not designed for flood control. The Tungabhadra Dam receives regular notification of floods in the Tunga and Bhadra dams (upstream dams) in advance, and outflows from the Tungabhadra Dam can be adjusted accordingly. So far, there has been no occurrence of floods exceeding the designated levels. The cost of compensation (for displacement) is budgeted into the cost of the projects; however, it is limited to those who are directly affected through submergence of land, and compensation is usually not provided for loss of livelihoods to fishermen or to downstream communities whose livelihoods are affected by the dams.

The middle and lower reaches of the basin are prone to droughts and one of the reasons for planning the irrigation uses of the Tungabhadra Dam was to provide water to the drought-prone region. However, experience shows that this has not worked as the designed cropping pattern was not adhered to. Another governmental measure is the allocation of funding to micro-watershed development and water conservation measures.

Demand for water for urban uses is increasing, particularly where regional towns are growing rapidly. Drinking water is supplied from the river and through other alternative sources such as groundwater and irrigation tanks to help meet the increasing demand. When there are drinking water shortages, the government provides water for domestic use (including drinking water) by bringing water via water tankers. Another strategy has been to drill bore wells especially for drinking water.

Data also shows that water-logging and salinization is a serious issue and has affected the total service area of the Tungabhadra by nearly 25 per cent (about 90,000ha of approximately 363,000ha). In the Tungabhadra, mining activities decrease the water depth and produce local iron contamination. This issue is going to be a major challenge in terms of technological options, requiring resources from reclamation and overall water management in the basin.

As already mentioned, in the Tagus River Basin, water abstraction for irrigation and urban supply, as well as industrial pollution emissions affect water quantity and quality; this sometimes compromises the terrestrial aquatic ecosystems. In Spain and Portugal potential water stress is related to water availability for the different sectors, especially because of the contrasting interactions between agriculture and urban water supplies. To mitigate water scarcity for these types of demand and to produce hydroelectricity power, large dams were built in the Tagus Basin with the total reservoir capacity of 11,140hm^3 (0.2hm^3 per square kilometre), in Spain, and 2750hm^3 (0.11hm^3 per square kilometre), in Portugal. In Spain a water transfer system from the Tagus River Basin to another basin (the Segura River) was implemented in 1978 with a yearly volume of transferred water that has not exceed 600hm^3 but that may increase to 1000hm^3 in the future, diminishing the total water availability in the Tagus downstream basin.

These pressures and impacts shape the conflicts between different water users in the river basins. In the Tungabhadra, both water quantity and quality create conflicts in inter-sector allocation. Moreover, the regulation between the upstream and downstream parts of the river basin and the consequent livelihoods of 10,000 fishing families create an additional cause of conflict. The

Tungabhadra Reservoir has also been losing water storage capacity and silting due to mining activities. In addition, soil erosion is also diminishing the reservoir capacity, causing conflicts between the two states, and there is a wide range of industrial activities with large plants and water demands. Irrigated agriculture is rapidly overtaking rain-fed agriculture, causing further depletion of reservoir capacities. The river is managed in each state by the irrigation or water resource departments of the state government, except in the case of the Tungabhadra Dam (for irrigation and hydropower), which is managed by the Tungabhadra Board, which was established in 1953 and functions under the Ministry of Water Resources of the Union Government of India. In tourist/religious places, transient populations have added to increasing consumption; but no data is available on water consumption. Similarly, there is no provision made to maintain minimum environmental flows.

In the Tagus Basin, regulation between the upstream and downstream parts of the basin (corresponding to the Spanish and the Portuguese parts, respectively) is governed by the Albufeira Convention, signed in 1998. The common factor in all conflicts in the basin is the depletion of water quantity and the decrease in water quality, as this impacts upon the terrestrial and aquatic ecosystem and compromises the different water users.

Methodology

In order to effectively manage land uses, a better understanding is needed between land use/land cover and water resources. The relationships between land and water resources can be characterized by indicator numbers, coefficients or indexes that describe the way in which a land system transforms the quantity and quality of water as it flows through a landscape. Some of those indexes are the runoff coefficient, which specifies the proportion of precipitation that becomes runoff for a given time period; the Soil Conservation Service curve number, which describes the capability of a soil to produce runoff according to its type, utilization and cover; and indexes related to pollutant loads, erosion rates or sediment loads. Regarding land-use changes, some of the short-term impacts are related to the increase of peak flood discharges, while the long-term impacts are related to changes in the average annual runoff.

In the TBSB, the analysis was based on qualitative and quantitative data collected through several methods, including structured and semi-structured interviews, focus group meetings, stakeholder workshops (see Chapter 3) and meetings with key informants. The data was analysed using several statistical tools and models to show the current trends and future scenarios. Scenarios were developed with stakeholder inputs to present the land-use options in the future and their impacts upon water use in the basin (see Chapter 4). Secondary data from various governmental departments were also compiled and analysed along with the primary data. In India, the available data across the sectors is not consistent, nor are basin-wide data compiled. At times, poor access to data in terms of both availability and quality made it difficult to run models.

Regarding the Portuguese sub-basin of the Tagus, based on the CORINE land cover maps of 1989 and 2000, a comparison was carried out with the aim of identifying land-use changes. In order to assess land-use changes in the whole Tagus River Basin, CORINE land cover maps from 1985 (for Portugal) and 1990 (for Spain) were compared with the equivalent maps from 2000, by means of geographic information systems (GIS). For the last decades, the generalization of the satellite imagery series allows for the assessment of spatial changes in vegetation cover. For example, a time series of the Normalized Difference Vegetation Index from the National Oceanic and Atmospheric Administration (NOAHH) sensor has been used to monitor the time evolution of vegetation cover in the central Pyrenees of Spain (Vicente-Serrano et al, 2004). A similar time series of annual NDVI values was created for the Tagus River Basin in order to assess the trends in the evolution of the vegetation cover during the period of 1982 to 2004.

SPSI in land- and water-use evaluations in the Tagus and Tungabhadra

In India, water is a state subject and the role of the central government is restricted to general water policy guidelines, planning and interstate matters. Interstate disputes are handled by tribunals set up under the Interstate Disputes Act. Since the TBSB is interstate by nature, disputes arose between the three states over the sharing of waters, which led to the Krishna Water Disputes Tribunal in 1969 under the Inter State Water Disputes Act of 1956 by the Government of India headed by R. S. Bachawat. The TBSB is part of the larger Krishna Basin and decision-making regarding it is part of the disputes proceedings. The Tungabhadra Board has been set up to implement the provisions of the Krishna Water Dispute Tribunal (KWDT) Award and the Bachawat Award, agreed upon during the mid 1990s, and stipulates a fixed scheme of water allocation in the Krishna Basin, including the TBSB. The operation of the Tungabhadra Reservoir is carried out by the interstate Tungabhadra Board and the award is currently under revision. As was evident at the stakeholder meetings conducted in the basin, the push and pull on the allocation of water within and between the states is significant. Neither the existing award, nor any other agreement, provides practical procedural arrangements for negotiating allocation and distribution under varying and changing circumstances.

The National Water Policy and the state water policies provide water-use priorities across different sectors, but do not contain much of relevance in terms of actual water-use planning and allocation. So far as sectoral balances are concerned, most of the changes in allocation take the form of changes from rural to another sector (i.e. addressing mainly changing irrigation use to other uses). In addition, there are no explicit legal agreements between sectors regarding the sharing of water. Competing water demands in the basin have to be met by reallocating water from other sectors. This creates a problem since reallocation would mean cutting down the water quotas from certain sectors and this could lead to conflicts. Water-use conflict in the Tungabhadra is a politically sensitive issue, leading to demonstrations by farmers and legal disputes between the states of Karnataka and Andhra Pradesh. On the whole, there is continuous pressure from different stakeholders and at different levels.

Stakeholder interactions in developing land- and water-use scenarios in the TBSB show how stakeholders perceive future developments, and the results of the model simulations shed light over three management actions. The scenarios included reasonable assumptions about water use and processing by industry, sewage treatment and sanitation measures by municipalities and local village councils (*panchayats*), and new cropping systems for different crops. The model simulation showed that:

- Climate change will require that erosion control should be a prime objective in landscape management and planning.
- Ongoing policies on sewage treatment in urban and rural areas have a significant impact in reducing nutrient loads to the river system.
- Introduction of new techniques such as SRI in rice cultivation plays an important role in saving water resources.

Figure 6.2 System of Rice Intensification (SRI) worker in the Tungabhadra Basin

Source: U.S. Nagothu

These are important interventions that need to be carefully considered in the basin.

In Spain and Portugal there are administrative systems for water management based on hydrological basin boundaries following the European Water Framework Directive (WFD). Regarding the Tagus River Basin, the authorities responsible for the management of the water resources are the Tagus Basin Authority in Spain, and the Tejo Hydrographic Region Administration in Portugal. According to the WFD, the basin administrations need to promote the Hydrographic Region Management Plans (HRMPs) that should provide support to the management, to the protection and to the environmental, social and economic valorisation of water resources – including the estuaries, the coastal areas and the aquifers, besides the surface waters. The first generation of management plans should be revised and updated by 2015. In both sub-basins of the Tagus River, the plans are prepared in a collaborative and consensual way through river basin councils. These forums (*fora*) facilitate the participation of the basin stakeholders and work according to the general national water strategies prepared by each country's national water councils. At the national level, water resources management is under the supervision of the Water Institute in Portugal and of the Water Authority in Spain.

Participation in land and water management

There are many dimensions and aspects to stakeholder participation in river basin management. Participation does not mean involving everybody in all decisions at all times, but rather thinking carefully about how to ensure that different interests can best be represented in different phases and forums of the multi-stakeholder process. The project examined the institutional framework, as well as the water policies and procedural rules, which form the legal basis for stakeholder participation in river basin management. Various conventions in Europe have provisions for public participation in river basin management or environmental impact procedures.

In India, during recent decades there has been a process of decentralization, partially in response to donor requirements, which brings in some scope for policy and stakeholder interface. There is provision for stakeholder participation in the National Water Policy of 2002 under clause 6.8 of the section on 'Planning' (p5): 'The involvement and participation of beneficiaries and other stakeholders should be encouraged right from the project planning stage itself', and in clause 12 of the section on 'Participatory approach to water resources management'.

In the context of our case basin, the TBSB, initiatives to promote participation and interventions to protect land degradation are taken up across different sectors and levels. Watershed development initiatives have focused on the holistic development of human resources, soil, and land and water management. A major intervention to protect the catchments and forest cover was the introduction of the 1980 Forest Conservation Act, which prevented the conversion of forest land for other purposes without prior approval. In addition, large areas within the Tungabhadra catchment were also declared as national parks to protect biodiversity. Despite such efforts, the catchments in TBSB are continuously exposed to a number of problems. Interaction between government organizations is limited. However, there are frequent project-level interactions, which to some extent act as sub-basin mechanisms.

Public participation in water resources planning and management in the Tagus River Basin is foreseen in legislation that resulted from the transposition of the European Water Framework Directive into the legal frame of legislation within Spain and Portugal. Both countries have achieved different levels of public participation in the water management process and have ratified several international conventions, which have provisions for access to information. Despite these issues, public participation has been relatively limited in terms of the decision-making process. The public participation process is based on three main general issues: information disclosure, public enquiry and active involvement of the stakeholders. Public involvement is accomplished by specific fora by means of meetings, paper brochures and advertisements in the newspapers, the internet and email. All information related to any water resource process is made available to the public.

Public participation and stakeholder consultations

In the TBSB there is currently no overall basin plan, or any public participation or multi-stakeholder organization, for creating or implementing river basin plans at state level; and there are no stakeholder consultations taking place in the Tungabhadra River Basin. Although there is interest in the concept of river basin organizations and various policy documents acknowledge river basin authorities or boards, little has moved on the ground. Despite the fact that there is no river basin organization at present, Karnataka has set up a corporation called Krishna Jal Bhagya Nigam primarily to generate (financial) resources from the public. Despite the lack of a multi-stakeholder platform or organization, there have been initiatives on the part of civil society and academic institutions especially, regarding issues of pollution. In many places there are also vibrant civil society initiatives on the equitable distribution of water, water rights and access, and issues of displacement and rehabilitation, as well as on micro-watershed development. Some budgetary provision is made for training programmes and exposure visits for farmers and officers, etc. The Water and Land Management Institutes are responsible for training and capacity-building. Some materials (literature) are also produced in local languages, although there is scope for improvement in terms of content and presentation. The irrigation departments have their own websites. In most of the states, each ministry or department comes out with a performance budget every year, which details operational and financial performance. Public environmental impact assessments are now compulsory; but very often, data and information are not given to those who oppose projects or those who are affected by projects. Consultations are also not held or informed consent taken. There is very little systematic and scientific public education provided on environmental and water issues. Mass media communication through the line departments is limited to publishing propaganda or campaign material, mainly in the form of printed posters, leaflets or resort to television/radio programmes. Folk art and traditional cultural forms, such as theatre and story-telling, are also used. Among the stakeholder workshops organized in the case basin, involvement and discussions were upgraded phase-wise across the different workshops through the involvement of stakeholders in the identification of the main issues, while actors within the water sectors (i.e. the officials from departments, researchers and NGO representatives) expressed their views based on their own experiences to be included in scenarios, thus giving scope to the SPSI interface.

Non-governmental organizations (NGOs) and civil society organizations (CSOs)

There are many NGOs (covering about 100 to 200 villages) working in the Tungabhadra Basin. In most instances, these NGOs are supported by international donor agencies. They take up a wide variety of activities and issues, such as education, drinking water and sanitation, watershed development, biodiversity (especially in areas within the basin that are part of the Western Ghats), organic farming, and so on. Currently, there is no umbrella organization that brings these NGOs together to share resources and to conduct training and workshops. The state Water and Land Management Institutes (WALMIs) are supposed to provide some of these inputs. Frequently, there are umbrella organizations at the state level that bring NGOs together within a state; but none are operational at river basin level.

Farmers' associations

There are many farmers' associations or organizations, as well as a few innovative farmers in the basin. In both Karnataka and Andhra Pradesh, participatory irrigation management (PIM), through the formation of Water User Associations (WUAs), has become official policy for the governments. However, in the Karnataka part of the basin the initiative to form WUAs is relatively recent; therefore, only a small number are operating at the moment, whereas in the case of the irrigated area in Andhra Pradesh the entire area is covered by WUAs. NGOs, in association with associations that promote organic farming, are prevalently involved in research and development (R&D), and in the training and promotion of low external-input sustainable agriculture (LEISA) practices and SRI paddy cultivation. The state agricultural universities also promote organic and sustainable agricultural practices, along with high-input non-organic practices. Similarly, the agricultural department through its extension network is also involved in the promotion of organic agricultural practices, although not very systematically.

Fisheries

The TBSB supports fisheries and livelihoods for a significant percentage of the population; however, there is no integrated approach to water management across departments. Pollution from industries leading to fish kill, along with illegal fishing, poor infrastructure facilities, dynamite operation, lack of training facilities, etc. have significantly affected this community. There are a number of policies and institutions already existing in TBSB that can facilitate the entry of youths, women and the poor into fisheries. The state fisheries department is engaged in research and development, demonstration, and outreach and extension activities. They also issue permits or licences for fishing in the larger water bodies, such as the dam reservoirs. Closer cooperation between national and regional government organizations and international and local NGOs is needed if planning of fisheries development is to be integrated with other sectoral development plans. This would also strengthen the capacity of organizations to plan and monitor databases at the local level.

Gender: Situation of women in the basin

There do not seem to be any specific women's groups working in the basin on the subject of water or water management. However there are self-help groups (SHGs) run for women on microfinancing, which are frequently promoted by different NGOs. In addition, there are *Mahila Mandals* (informal women's groups/associations at village level). There have also been attempts to federate the SHGs at district levels. Nevertheless, there is no organized effort to mobilize or organize women on the issue of water or to give voice to their demands and interests. Indian society is, by and large, patriarchal, and women's participation in government and decision-making bodies tends to be limited. How patriarchy manifests and operates differs from state to state and is also culturally embedded. However, over the years, there has been a slow change and women have been increasingly gaining access to government bodies and elected bodies at different levels. This change has been due, in part, to the spread of education and to the efforts of social movements and women's organizations and NGOs to build awareness and to organize women. Initiatives from the state to reserve up to 33 per cent of elected bodies (up to district level) and public institutions such as co-operative societies, etc. for women's participation have also had a hand in this change. Gender inequity, however, differs in the various social groups and, more often than not, women do not have access to institutional credit.

Stakeholder access to information and decision-making

Concerning access to information, the Tungabhadra does not have a basin-wide plan; hence, interest groups have to rely on the detailed project reports (DPRs) of individual projects, which mainly outline project design, cost estimates and planned water use. These documents do not usually include governance issues, including stakeholder involvement. People frequently do not have access to data and information; furthermore, making data available to different stakeholders in an understandable form has been generally weak. Often there is no consistency in data sets maintained by different agencies, which causes difficulties in negotiations concerning interstate rivers. A commonly agreed data set is a key issue for stakeholder involvement too. The 2002 Karnataka State Water Policy states that a data information centre would be set up and data protocols developed to remedy the situation. The situation has also slightly improved with the Right to Information Act now in place in most states.

In the Tagus River Basin, development and elaboration of the hydrological plan requires considerable participation by the public, in general, and by the stakeholders, in particular. Besides the water users (with emphasis on the agricultural and industry sectors), other relevant stakeholders are the water managers, social networks, private companies and NGOs.

In the TBSB, public officials generally dominate the decision-making process. Engineers and multi-stakeholder fora are, most notably, not in place, although Water User Associations, Water Development Committees and self-help groups provide some opportunities for civil society participation, with financial support provided for their attendance at meetings. But they play a very limited role as the officers of the respective departments make most of the decisions, which are often taken with political interests in mind. Issues related to a particular WUA are discussed at a local level; but issues of policy, etc. are discussed at the state capital. The public are not involved in problem identification, and although public hearings are compulsory for clearing projects with environmental- and displacement-related impacts, the experience of many stakeholders and civil society organizations has not been encouraging, as these public hearings are often manipulated to suit the interests of the proponents of the project. The hearings often do not have any influence on the final outcome. Water policy documents and legislation do contain clauses to support stakeholder participation, such as the Participatory Irrigation Management Act and the 2002 National Water Policy, which mentions stakeholders under clause 6.8 of the section 'Planning 41' and in clause 12 of the section on 'Participatory approach to water resources management'. In the 2002 Karnataka State Water Policy, the word stakeholder appears in the context of participatory irrigation management and water users' associations to manage irrigation water. Although the 2002 National Water Policy and the Karnataka State Water Policy (Andhra Pradesh has yet to come out with its water policy) do mention stakeholder involvement, very little has been done in reality. The only area where involvement of the stakeholders is sought is in the area of irrigation water management as part of sectoral reforms; thus almost all documents mention water user participation for irrigation water. In the drinking water sector, efforts are being made to involve users in managing drinking water schemes, both in urban and rural areas. Micro-watershed development is a major programme in rural areas, funded by the Ministry of Rural Development, the Ministry of Agriculture and multilateral and bilateral agencies. With guidelines for community participation in programme implementation, institutions such as village councils, a Watershed Development Committee and self-help groups have been set up and include the poor and women. Thus, in all three major sectors – irrigation, drinking water and watershed development – efforts are being made to incorporate participation. There is no consultation amongst the different stakeholders on the question of inter-sectoral water allocation leading to conflicts. However, the 2002 National Water Policy does provide for the establishment of river basin organizations for the development and management of a river basin, as a whole, or sub-basins wherever necessary. Various studies also indicate that there is a large gap between the official/policy rhetoric and practice on the ground, thus indicating the need for a strong development of the SPSI in practice.

Concerning institutional, policy and legal frameworks, India has a federal system of governance and the constitution has categorized all subjects of governance into three categories – namely, central (or union) list, state list and concurrent list depending on who (union, state or both) has the power to legislate on that particular subject or who has jurisdiction over that subject. Water is a subject that comes under the state list, meaning that each state can legislate as well as formulate laws, rules and regulations with regard to water. The only exception is in the case of inter-state (transboundary) rivers where the union can step in – especially if there is an inter-state dispute on sharing water – and appoint a tribunal that would decide on the allocations. This is done through the 1956 Inter-State Water Disputes Act. As mentioned earlier, the Tungabhadra River is a tributary of the Krishna River, and water utilization is governed by the Krishna Water Disputes Tribunal.

Water for agriculture is highly organized. Over the last few years there have been moves towards increasing water charges in the basin. However, no special provisions have been made to assist poor farmers, and rich farmers often capture the subsidies. So the viewpoint is that the basis of pricing should be shifted from the present practice of crop-area basis to volume of water consumed. There is also a suggestion that differential tariffs should be introduced. Users who consume above a particular level should be charged higher prices per unit. This is also reflected in the 1992 Irrigation Committee Report (known as the Vaidyanathan Committee Report). Water for industrial use is charged at a much higher rate compared to use by farmers – thus, there is a cross-subsidy. During the last few decades, there has been a process of water reform leading to changes at the national and state level after the adoption of the first National Water Policy. This reform has been partially driven by increasing water scarcity and, in part, by conditions put forward by donors based on the Dublin Principles. This has led to water being viewed not only as a social good, but also an economic good, taking into account the need for both water conservation and cost recovery. The main thrust of water sector reforms was to transform the role of the government by transferring part of the existing governmental management functions to users and private actors, including the transfer of operation, maintenance, management and collection of water charges to user groups. This was meant to foster a sense of ownership at the user level. A second thrust of the reforms is to set up new bodies, at local and state level, to take over part of the functions of the government, such as Water User Associations, to locally manage irrigation schemes which also includes the much more broad-ranging setting-up of new water regulatory bodies. The reduction of the role of the state in the water sector is also linked to the promotion of incentives to ensure that water is used more efficiently. The main consequence of this change is the call for private-sector involvement in water control and use, from planning to development and administration of water resource projects.

The focus on participation is usually at the 'tail end' of the process; hence, reforms provide no great possibility for farmers and users to participate in making decisions that affect them. This can lead to the blanket imposition of new systems of local water use and control schemes based on commercial principles, even where there may be successful systems of water governance already in place. The linkage within the two states between land and water rights means that the reforms are likely to reinforce inequities in access to water between landowners and landless people. The decentralization process strengthened the *panchayati raj* system (local institutions with a three-tier district system), consisting from the highest level down of the *panchyat* (or *Zilla Parishad*), the *taluka panchayat* and, at the lowest level, the *gram* (village) *panchayat*. About 18 subjects under the state list were transferred to the *gram panchayats*, including the village water bodies (village tanks and ponds) as well as drinking water and sanitation, watershed development, etc. There are also efforts to strengthen direct democracy at the village level, as it has been made compulsory to convene the *gram sabha* (village assembly of all adult members) at least twice a year and all important decisions are supposed to be taken in the village assembly. Of course, these processes are all on paper, and practice has not kept pace with the expectations generated through these legislative measures.

At present, it is the government departments (such as the water resource departments) from the two states that make decisions. Of course, the Tungabhadra Board, which functions under the Ministry of Water Resources of the Government of India, is also an important player. Apart from these, local politicians (such as members of parliament and members of the legislative assemblies) – that is, state- and district-level politicians – exert pressure on the system. Another important set of players are the influential industries. There have been a number of civil society movements against industrial pollution and because of the pressure they exerted, certain monitoring systems are now in place. There are no major civil society initiatives that organize the farmers or other stakeholders on the issues of river basin management.

In the Tungabhadra, pollution control is separated institutionally from abstraction management. In Karnataka, the State Pollution Control Board for Prevention and Control of Water Pollution is responsible for developing a pollution management policy, and its regional offices are charged with implementing pollution control. Water abstraction is controlled by the state Department of Water Resources in the first instance, although this body is responsible for approving or rejecting applications for water use from relevant sectoral

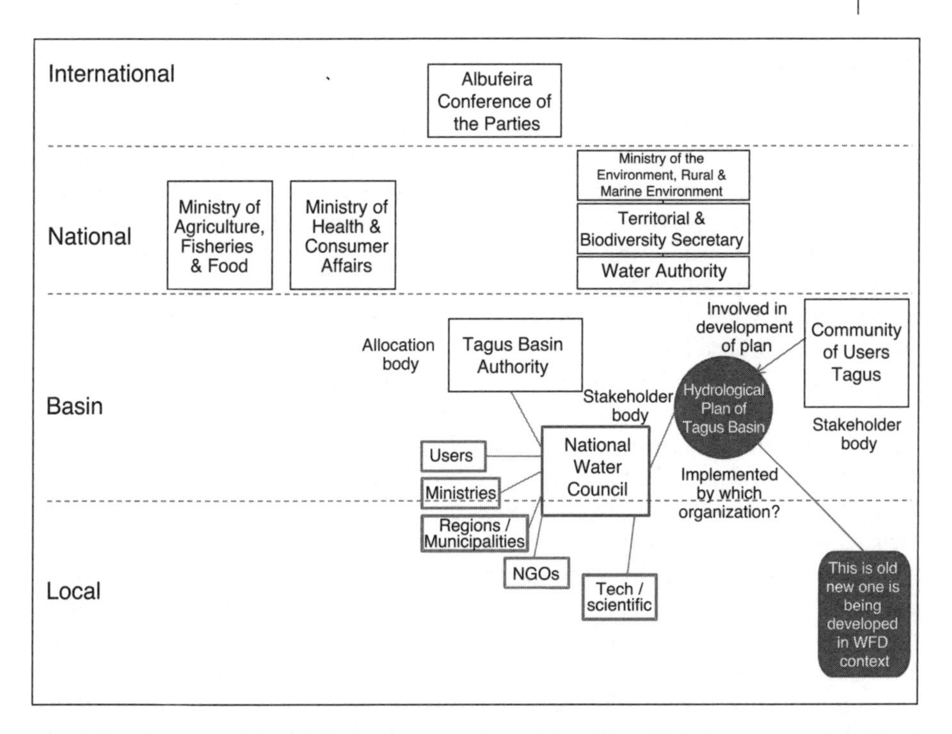

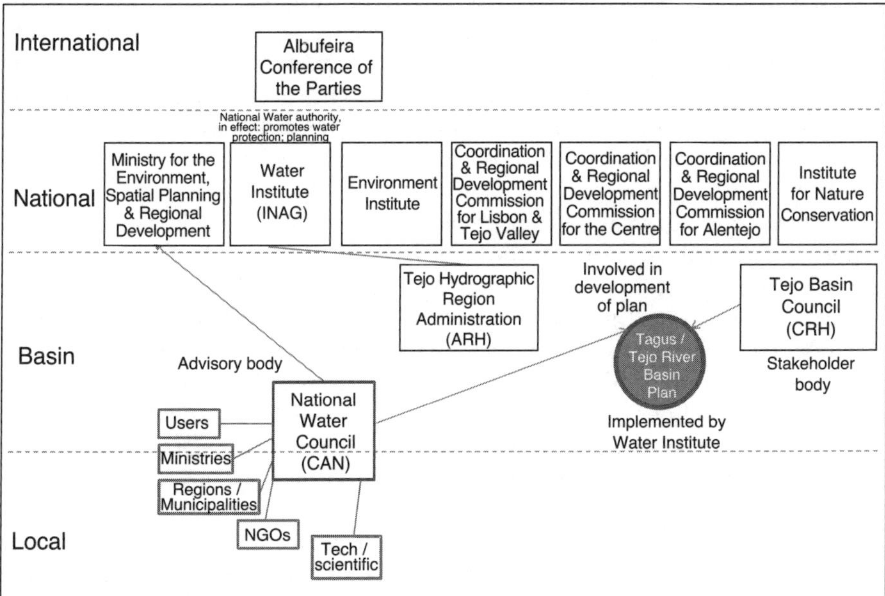

Figure 6.3 Spanish (top) and Portuguese (bottom) institutional structure organograms

Source: Campbell, D., Rieu-Clarke, A., Allan, A., Gooch, G.D., Stålnacke, P. and Nagothu U.S. 2009. Final Stakeholder Analysis - Stakeholder Participation in the STRIVER Basins. STRIVER Report D4.2. 81p. http://kvina.niva.no/striver/Disseminationofresults/STRIVERReports/tabid/80/Default.aspx

ministerial departments. It is worth noting that this department exists only in Karnataka thus far, replacing the former Department of Irrigation. This name change has not yet occurred in Andhra Pradesh. However, questions need to be raised as to whether or not departments responsible for irrigation can change quickly into departments responsible for all aspects of water use in the context of IWRM without favouring agriculture, at least initially. The Tungabhadra Board, which reports to the union-level Ministry of Water Resources, controls water allocations and power output from the Tungabhadra Dam to the basin states; but it is not a licensing body. Water management is therefore managed not at the basin level, but at state level.

There is no river basin committee or council; hence, there is no question of coordination between different line departments at a basin level. However, there are several departments across the basin addressing various management aspects such as agriculture, fishing, pollution, animal husbandry, forests, the Tungabhadra Board, etc., while some of the departments have set up their own committees in which there is representation by the public (apart from the governmental and departmental representatives). Civil society groups are active in certain areas, such as supporting the WUAs and pollution control. As mentioned earlier, training sessions, extension services concerning water management, crop water requirements, etc. are provided by WALMIs. Hands-on training is provided by the Command Area Development Authority (CADA). The WALMIs and irrigation departments have developed some material (for training, campaigns, etc.) in the local languages. However, training is very sporadic, although systematically organized. Nowadays, the focus of training programmes is PIM and the office bearers of WUAs go through some training, particularly regarding the procedural and management issues related to WUAs. In places where the NGOs are active, there is focused training and capacity-building.

In Spain and Portugal, as in all countries that belong to the European Union, water management is based on the river basin (as territorial units) and on the administrative structure defined by the Water Framework Directive according to the legislation that transposed the directive into each country's legal framework. The river basin agencies have, among others, three main roles: licensing, surveillance and environmental protection. Both Portugal and Spain apply the user pays and the polluter pays principles based on real costs and benefits. New water saving technologies and new types of crops are being introduced in order to improve the efficiency of the distribution systems and to reduce water consumption by the agriculture sector.

Research on land-use change and land and water interactions is currently conducted at universities and at research centres within the European Union (EU) framework and national research programmes. Scientists provide results to national and regional water managers through advice and consulting activities for the government, basin agencies and institutional stakeholders, as well as through NGOs and participation at national and basin councils.

Conclusions

The chapter set out with the proposition that land-use changes not only have a major impact upon water resources, but also have great potential for modifying the hydrological cycle within river basins and very often the relationship between land use, water quantity and quality. The study of both the Tungabhadra Sub-Basin and the Tagus Basin validate the position that land-use change, coupled with other macro-level policy shifts in water-use prioritization, allocation and competing demands by different uses and users, is an important determinant and driver in controlling water use and demand. In the TBSB, rapid urbanization, population growth, the rapid rise of high-input intensive (irrigated) agriculture, and a rapid increase in mining activity in the region have contributed to high rates of siltation, decrease of capacity of the Tungabhadra Dam and a rise in pollutant levels, especially during the summer season. There has also been an increase in forested areas, especially in the upper reaches of the basin, thus converting some of the 'blue water' to 'green water'. Lisbon and Madrid, the capital cities of Portugal and Spain, respectively, both lie within the Tagus Basin and their expansion and growth have placed heavy demands on water resources. Rain-fed agriculture is decreasing rapidly and giving way either to natural vegetation and regeneration or to irrigated agriculture. It has been found that water returned to the river system from irrigation, urban habitation and industries has had adverse impacts upon water quantity and quality, sometimes seriously affecting terrestrial aquatic ecosystems. Furthermore, in both the basins, new demands for water have emerged necessitating reallocation of water across different uses and users, leading to conflicts across and within sectors, affecting economic and environmental prerogatives.

However, land use is determined by activities that usually fall under the purview of many departments other than those dealing with water resources. In addition, there are problems associated with the rivers being transboundary (international or interstate). Since water in India is a state subject, and not a union subject, this means that both states have different kinds of laws and institutional structures operating on the ground. Thus, if it is acknowledged that land-use changes, both in terms of what use it is put to and how, are an important determinant of river flow in terms of both quantity and quality, good governance in this respect would require a closer collaboration between different departments within the state or country, as well as between countries.

In the Tagus Basin, the Albufeira Convention and the Water Framework Directive have proven to be two important instruments in bringing these concerns together under a single process. Basin level authorities have been set on both sides and basin plans are prepared by an iterative process of consultation in which scientists as well as stakeholders also play a part. In this sense, SPSI processes are an important component in basin planning. However, land use does not seem to figure directly as an issue in these processes. Nevertheless, since the basin plans are arrived at through a process of bilateral consultations, many land-use implications are addressed. In India, the Tungabhadra Board is the only collaborative body involving the two riparian states of the Tungabhadra River. However, the board has a very limited mandate: that of looking after the allocation of water from the Tungabhadra Dam between Andhra Pradesh and Karnataka, according to the Tribunal Award. It does not have a basin-wide function. All of this calls for a better science–policy–stakeholder interaction so that scientific and socially acceptable land-use and water allocation policies can be put in place. However, SPSI processes are not institutionalized at the basin level within states or between states. Since rapid changes in land use are creating a serious impact upon river flow quantity and quality, it is important that land-use planning in the basin is conducted collaboratively by the two states. This necessitates the setting-up of basin-wide institutions within the two states, and institutionalizing collaborative planning for the two basins on the basis of a common framework such as the Water Framework Directive, but adapted to the conditions of the Tungabhadra Sub-Basin.

References

Beguería, S., López-Moreno, J. I., Lorente, A., Seeger, M. and García-Ruiz, J. M. (2003) 'Assessing the effect of climate oscillations and land-use changes on streamflow in the Central Spanish Pyrenees', *Ambio,* vol 32, no 4, pp283–286

Beguería, S., Raju, K. V., Manasi. S., Latha, N., Umesh Babu M.S. and Nagothu, U.S. (2008) Land Use and Land Use Change - Implications for Water Resources and Water Use in the Tagus and Tungabhadra Basins. STRIVER Task Report No. 9.3 http://kvina.niva.no/striver/Portals/0/documents/STRIVER_D9_1Task_Report_9_3.pdf, last accessed March 2010

Beguería, S., López-Moreno, J. I., Gómez-Villar, A., Rubio, V., Lana-Renault, N. and García-Ruiz, J. M. (2006) 'Fluvial adjustments to soil erosion and plant cover changes in the Central Spanish Pyrenees', *Geografiska Annaler series A – Physical Geography*, vol 88A, no 3, pp177–186

Bosch, J. M. and Hewlett, J. D. (1982) 'A review of catchment experiments to determine the effect of vegetation on water yield and evapotranspiration', *Journal of Hydrology,* vol 55, pp3–23

Gallart, F. and Llorens, P. (2001) 'Efectos de los cambios de uso y cubierta del suelo en los aportes del rio Ebro y su evolucion futura', in N. Prat and C. Ibañez (eds) *El Curso inferior del Ebroy su delta*, U. Cantabria–U. BarcelonaSantander, pp51–57

García-Ruiz, J. M. and Lasanta-Martinez, T. (1990) 'Land-use changes in the Spanish Pyrenees', *Mountain Research and Development*, vol 10, no 3, pp267–279

Gooch, G. D. (2006) 'Actor network theory as a tool for analyses of multi-level water governance', Presented to the International Workshop on Governance and the Global Water System: Institutions, Actors, Scales of Water Management Facing the Challenges of Global Change, Global Water System Project, Bonn, Germany, 20–23 June 2006

Government of Karnataka *District at a Glance Documents: 1975 to 2005*, District Statistical Office, Government of Karnataka, India

MIMAM (1998) *Libro Blanco del Agua en España*, Ministerio de Medio Ambiente, Madrid

MIMAM. (2000) *Plan Hidrológico Nacional: Análisis de los sistemas hidráulicos*, Ministerio de Medio Ambiente, Madrid

Sahin, V. and Hall, M. J. (1996) 'The effects of afforestation and deforestation on water yields', *Journal of Hydrology*, vol 178, pp 293–309

Vincente-Serrano, S., Lasanta, T. and Romo, A. (2004) 'Analysis of spatial and temporal evolution of vegetation cover in the Spanish Central Pyrenees: Role of human management', *Environmental Management*, vol 34 no 6, pp802–818

Confederación Hidrológica del Tajo (2001) Plan Hidrológico de la cuenca del Tajo, Confederación Hidrográfica del Tajo (CHT), Madrid

Karnataka at a Glance: 1992–1993, Directorate of Economics and Statistics, Bangalore

7

The Science–Policy–Stakeholder Interface and Environmental Flow

Dag Berge, David N. Barton, Dang Thi Kim Nhung and Ingrid Nesheim

Introduction

Environmental flow is an important measure to mitigate the negative impacts of hydropower regulation in watercourses. Initially, the regulation of rivers often left them dry downstream of dams or diversion points. However, during the 100-year history of hydropower regulation, it was recognized early on that a minimum release of water was necessary to protect a river's ecology, as well as to provide adequate water for other uses. Power companies and authorities were originally reluctant to introduce the concept of minimum release as it reduced power production. Due to the development of environmental management authorities in the Western world during the 1970s, minimum release, however, became increasingly common in hydropower regulations. This minimum release normally consisted of a single flow value that was released throughout the year. Thereafter, the idea of minimum flow was further developed to typically include two flow values: a low value during winter and a higher value during summer. During the latter part of the 20th century it became obvious that both the river environment and other interests associated with water use in a particular river system could benefit considerably by adjusting the minimum release more in accordance with the actual need. This gave a much more variable minimum release than before. The term environmental flow was therefore born.

The STRIVER project work on environmental flow has been fourfold – namely:

1 Review the international methodology regarding environmental flow assessment.
2 Review the concessions of the 56 hydropower regulations in the Glomma River's 100-year-old hydropower history with respect to methodology applied to assess minimum releases in the different regulations.

3 Assess which of these methods could be used in the new hydropower development scheme that is evolving in the Sesan River within Vietnam and Cambodia.
4 Elaborate upon an assessment methodology based on the relationship between the pressure and the impact which hydropower regulation exerts on river ecology and different water use in line with the principles of the European Union Water Framework Directive (WFD).

Review of environmental flow methods

Internationally, more than 200 methods to assess environmental flows in regulated rivers are described (Tharme, 2003; Halleraker and Harby, 2006). It would be an impossible task to go through all of these methods singly, so they have to be treated in a group-wise fashion, as applied by several earlier reviewers (Jowet, 1997; Dunbar et al, 1998; Tharme, 2003; Scruton et al, 2005; Halleraker and Harby, 2006) as follows:

- hydrological methods:
  - hydrological reference table method;
  - identification of central hydrological events;
- hydraulic methods;
- functional connections between physical alterations and river biology;
- holistic methods;
- hybrid model framework.

Each of the groups will have methods that involve both mathematical model simulations and subjective evaluations based on expert judgements.

The principle underlying most of these hydrological methods is to find an acceptable minimum flow. This is provided as a percentage of the natural flow (i.e. usually as a percentage of mean annual flow). In order to be able to assess this daily flow, measurements are needed spanning several years. If such data do not exist, it is possible to perform calculations via modelling or by proportional scaling of the measurements available from a river situated nearby.

The hydraulic group of methods was popular during the 1970s. The methods describe, via hydraulic models, how different water flows affect an area of the river bed covered by water, as well as water velocity, sedimentation, erosion, etc. (Halleraker and Harby, 2006). The hydraulic methods provide more detailed and localized information on how regulation will impact upon the physical environment in comparison to hydrological methods (King et al, 1999); but they do not include any functional connection between the physical changes and the preferences of local flora and fauna. Such studies include habitat requirements of different species, temperature preferences, water velocity, growth relationships, etc. The impact upon the ecology of a river through physical alterations can be studied and quantified, and available data can be put into predictive models, which can describe how a certain regulation will affect local biology.

Holistic methods take into account the flow needs of the river biology and human water user interests. The concept is a structured evaluation of composite expert judgements, where experts from different disciplines (expert panels) work together in interdisciplinary workshops. At least 16 such methods (Tharme, 2003) exist. So far, the methods have only been used extensively in Australia and South Africa (Halleraker and Harby, 2006); but Dunbar et al (1998) recommend that they be further developed and adapted to British and other conditions.

The hybrid model frameworks consist of model frameworks that link the former four categories together (hydrological, hydraulic and habitat models and holistic methods). The models can be used in estimating the environmental impacts of different regulation manoeuvring rules (e.g. the impacts upon water quality, fish, etc.). The problem with these model frameworks is that they are too complex to really be used in real-life conditions to any great extent.

The Glomma: Hydropower regulation and the application of environmental flow

History

Hydropower production is an important water use in the Glomma River and has existed for more than 100 years. In the Glomma River Basin there are 56 hydropower stations and 26 hydropower reservoirs. Coordinating the manoeuvring of regulations is taken care of by a water management association, the Glommens og Laagens Brukseierforening (GLB), among the owners of the different hydropower stations. The GLB has, to date, 18 power companies as members and performs several water management responsibilities in the basin, which include ensuring that the concession conditions with respect to minimum flow in rivers and water levels in reservoirs are not violated. The GLB was established in 1918 and the association also manages the hydrological gauging stations in the river basin (water flows and water levels).

Most hydropower regulations in the Glomma River Basin (if we omit the pure run of the river regulations and the oldest regulations) stipulate some kind of minimum release, compensation flows or environmentally motivated rules for water-level variations in reservoirs. The way in which this is carried out can be roughly described as expert judgement based on baseline studies of ecological items and water-use items. In Norway this is often referred to as the expert panel method. Under the old regulations there was often only one expert allocated to take care of environmental aspects, and this was usually the regional fish inspector belonging to the Directorate for Nature Management. In more recent regulations there are several experts and river users involved in assessing the minimum water flows and the water levels. This expert group is often called an expert panel. The experts included in this panel may vary from case to case, and there is no clear methodology upon which they base their judgements, which may cause inconsistencies in the results generated. For example, when a concession is up for renewal, it is not always easy to see how the panel arrived at a certain compensation flow as their methods are neither transparent nor replicable.

With reference to international terminology, the expert panel method is used in Glomma to assess minimum flows, or compensation flows, or water levels in connection with hydropower regulations. There is, however, a need to conduct these exercises in a more structured way with respect to which type of experts and stakeholders should be involved and which ecological values and which river use values should be included. In addition, a more structured, quantifiable, replicable and transparent methodology should be applied in order to achieve a certain water flow and associated water-level manoeuvring rules.

The formal process for hydropower concession in Norway

The formal process for river regulation in Norway is dependent on the size of the project. If the hydropower project is 40GWh or more, then the project comes under the Plan and Building Law (PBL). This stipulates an environmental impact assessment (EIA), a social impact assessment (SIA) and public and stakeholder involvement and hearings, etc. If a project is between 30GWh and 40GWh, the Norwegian Water and Energy Directorate (NVE) decides if treatment in accordance with PBL is necessary or not. If the project is below 30GWh, the project does not need to be treated under the PBL and the treatment process is much simpler. For these smaller regulations there is a general requirement, after the Water Resources Law, that their minimum flow is at least the size of 'common low flow' (approximately 10 per cent of average flow). In the larger projects, the question of minimum flow is taken care of during the application process for licensing and no specific requirements for the size of the flow are provided in the legal system.

In larger projects (>40GWh), such as the Glomma River, the applicant (hydropower company) provide an announcement document (in Norwegian, *Melding*), including a description of the planned project, as well as a programme for impact assessment. The NVE then arranges a public meeting with representatives of all relevant stakeholders (local authorities, water users, landowners, rights holders, NGOs, etc.). The meeting is held in the municipality that is most affected by the project and is open to all. The announcement document is sent out for public hearing four weeks in advance and is open to comment. Based on the results of the hearing, the NVE presents a detailed programme for the impact assessment which the applicant has to conduct before the application can be taken any further. After the impact assessment is conducted, the applicant updates the announcement document to include the recommended mitigation measures, etc. The announcement document is then given application status, which is sent to the NVE, who then sends the application out for public hearing for three months. The NVE then evaluates the project and provides a recommendation (positive or negative) and sends it to the Ministry of Petroleum and Energy (OED). The OED conveys the recommendation on a limited hearing (to other relevant ministries, directorates, municipalities, etc.). The OED subsequently makes a proposal for a decision and sends it to the King in Council for a final decision. In addition to being assessed under the

PBL, large projects are also treated under the Water Resources Law and the Watercourse Regulation Law. A requirement for environmental flow is normally in accordance with the requirements of abatement measures in new licences.

Today, only publicly owned companies can obtain a licence, and the licence then has no time limit. Until recently, however, private companies could also obtain licences; but these were limited to 50 years with reversion to the state after that time. The private company could, however, then buy the hydropower plant and regulation back, and apply for a new licence. Licence conditions, rights and obligations of the licence and rules of operation are all significant. After 30 years, these conditions can be revised. In a few cases, in order to test out abatements, such as environmental flow, a licence has been given for a test period of five to ten years, after which some abatement measures are adjusted if found appropriate. When the licence elapses after a period of 50 years, the power company has to apply for renewal of the licence. Normally, this entails only some small adjustments of the conditions, such as implementing environmental flow in earlier dry stretches. A hydropower regulation licence has never yet been withdrawn in Norway.

The PIMCEFA method for environmental flow assessment

Environmental flow has mostly been used in connection with assessing minimum flows in river stretches that would have been dry after regulation (i.e. downstream of dams, diversion points, etc.). However, regulation creates many other major changes in a river, which requires a broad spectrum of abatement measures and diverse ways of thinking when it comes to environmental flows.

For example, the cascade development of hydropower dams in the Sesan River in Vietnam/Cambodia (see Figure 7.1) has changed the river from a continuous water body to a cascade of lakes. Regarding fish production for local livelihoods, as well as for other local water uses, it is the lakes that will be the most important water bodies in the future. Thus, water-level management of reservoirs will be more important than the minimum flow in the river stretches between dams. The only way in which these can be regulated is through water release from the reservoirs, either via the turbines, the spillway or the bottom valves. Thus, knowledge of environmental water release is necessary so that hydropower companies can initiate and achieve environmental flow.

In our work, we therefore adopted the following definition for environmental flow:

> Adopting water release manoeuvring rules for the different reservoirs to obtain as favourable water levels (and water flows) as possible for the total river ecology and the human water use interests, within the constraints set by the economical feasibility of the regulation. This applies both for reservoirs and river stretches.

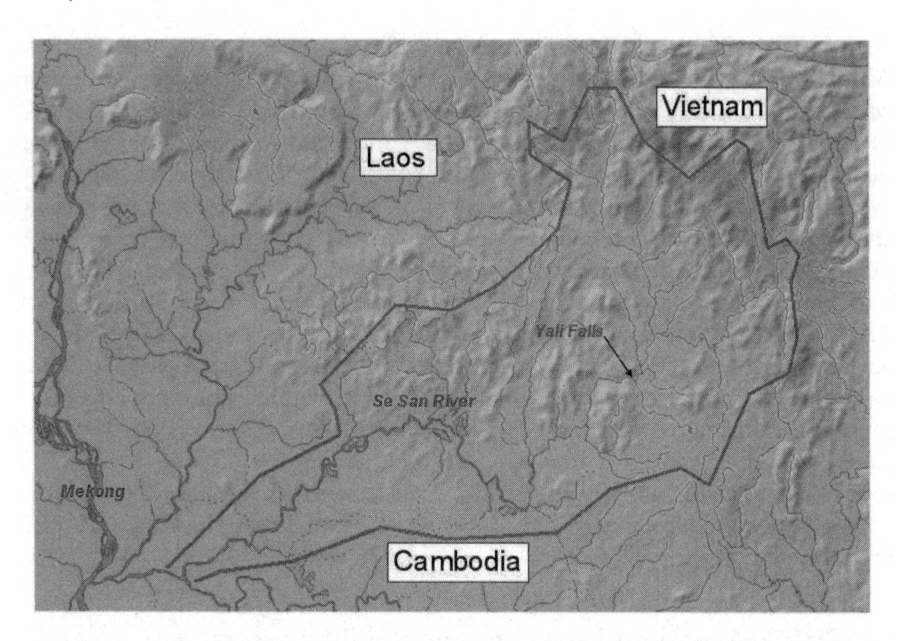

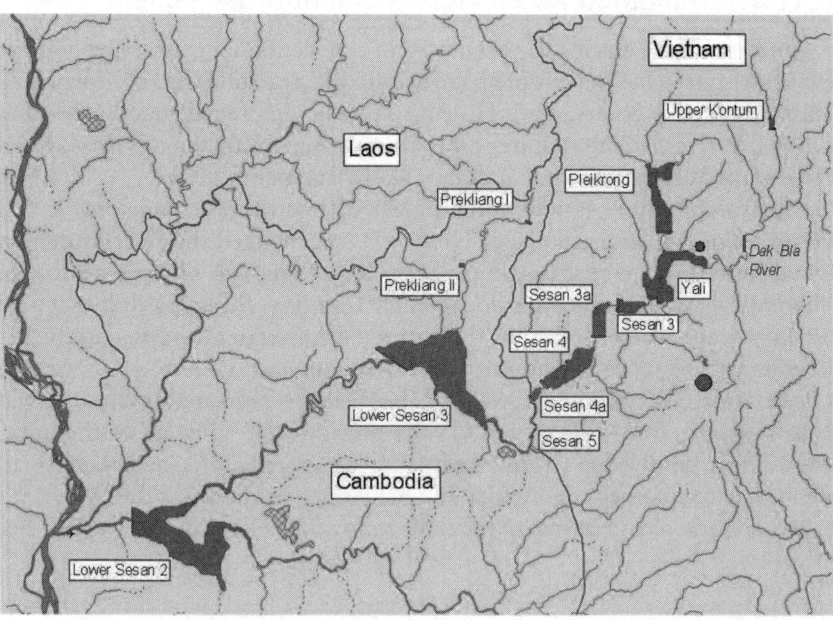

Figure 7.1 Before 1998, the Sesan River was a continuous river where fish could migrate all the way from the Mekong and far up in Vietnam (upper panel); the regulation scheme transforms the river to a cascade of lakes (lower panel)

Source: Re-drawn by D. Berge based upon an economical map obtained from Electricity of Vietnam and maps obtained from Demis Web Map Server (www.demis.nl)

The review of Glomma hydropower regulations revealed that expert judgement, often conducted by a loosely defined expert panel, was the most usual method of assessing minimum flow. According to the GLB, as well as the Norwegian Directorate for Water Resources and Energy, the expert judgement method will always be central in the process of making decisions about minimum releases in Norway. However, the GLB also considers that there is a great need for making the method more structured. In international terms, the expert panel method belongs to the holistic type of methods (King and Louw, 1998).

Inspired by the experience gained from the most recent environmental flow assessment case in Glomma (i.e. the renewal of the concession of the Øyeren Regulation; Berge et al 2002), the Pressure Impact Multi-Criteria Environmental Flow Analysis (PIMCEFA) method was elaborated upon as part of the STRIVER project. The method can be used to design environmental flow, as well as to assess the degree of damage done to different river values at different levels of regulation.

In this method, an expert panel consisting of local experts (fishermen, farmers, boaters, etc.) and professional experts, both on key ecological elements (ecological values) and key water use elements (water use values), is appointed for each of the different river sections that are undergoing evaluation in the context of environmental flow.

Members of the expert panel try to construct the optimum water level curve over the year for the different water values (ecological values such as water quality, water vegetation, bottom animals, fish, etc., and user values such as drinking water, transportation, irrigation, fishery, etc.) based on local knowledge and baseline studies combined with professional expert judgement. The panel members then go on to identify the critical periods for the river values that they represent (i.e. periods where certain water levels have to be kept). From the different optimum curves a preliminary resultant curve can then be structured. The curves shown in Figure 7.2 are from the work conducted during the renewal of the Øyeren Regulation Concession in the Glomma River (Berge et al, 2002).

Once the critical periods for the water value (ecological value or user value) are identified, the next step is to evaluate how seriously the different river values are affected by the regulation. A semi-quantifiable model, called a pressure-impact curve, should be created. To illustrate this point, the river value 'fish production' in a regulation which includes river diversion is used as an example (see Figure 7.3). If all the water is taken and the river ends up dry, then fish production is damaged by 100 per cent. If no water is taken (no regulation), the damage to fish production is 0 (zero). The simplest model to demonstrate these two points is the straight line (see Figure 7.3). However, this model can be improved considerably by expert judgement (as well as in cases where few data exist). We know that most fish species can adapt to a 30 per cent year-to-year variation in water flow, so to divert up to 30 per cent of the water is not considered a very serious threat to fish stocks. At the other end of the scale, when most of the water is diverted and fish stocks are depleted, there will only be small pools of water (e.g. between the stones on the bottom), producing the

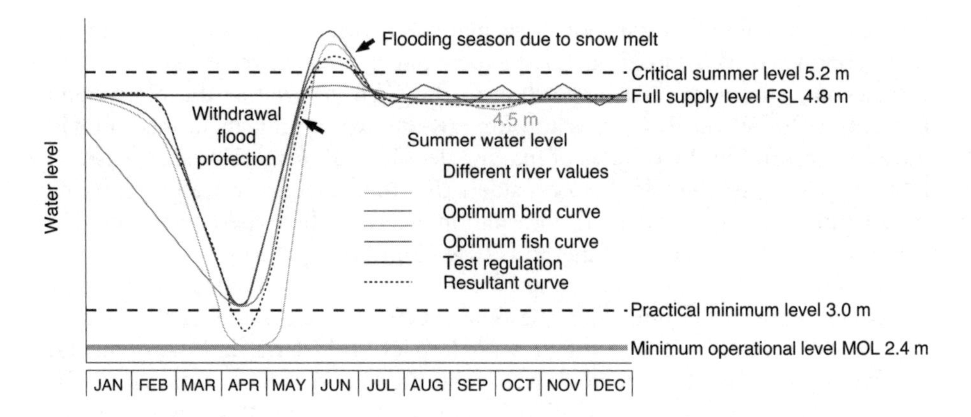

Figure 7.2 Optimum water-level curves for the different river values in the Glomma River Delta Northern Øyeren Nature Reserve

Source: Data derived from the renewal of the Øyeren Regulation Concession (Berge et al, 2002)

same negative effect. Most ecosystem damage follows such a sigmoid curve, while user interests often follow a curve as noted in Figure 7.3.

To help in evaluating the different impacts against each other, a multi-criteria analytical tool (MCA software) was used in the project. From the many software products available, we chose a commercially available software package called DEFINITE (Janssen and Herwijnen, 2007) to conduct multiple-criteria and benefit-cost analysis. We used DEFINITE in PIMCEFA as a method of documenting stakeholder and expert judgement and as a tool for ranking alternative environmental flow levels under consideration. A particularly useful feature of the DEFINITE software in the context of PIMCEFA, and compared to other commercially available packages, is the function that lets the user define pressure-impact curves manually using any functional form. This is a crucial advantage, as PIMCEFA relies on being able to accurately capture expert knowledge regarding the link between river flow and impacts upon ecological and user interests in the form of pressure-impact curves. Pressure-impact curves are otherwise known as 'value functions' in the MCA literature (Beinat, 1997).

In Figures 7.4 and 7.5 we have illustrated the main steps in using MCA software such as DEFINITE to rank river flow alternatives. The following only refers to the steps relating to DEFINITE in the PIMCEFA approach:

- Step 1: define the alternative river flow levels under consideration.
- Step 2: define the hierarchy of impact indicators.
- Step 3: convert hand-drawn pressure-impact curves to 'value functions' in DEFINITE.
- Step 4: elicit relative weights for impact criteria from stakeholders (or set equal default weights).

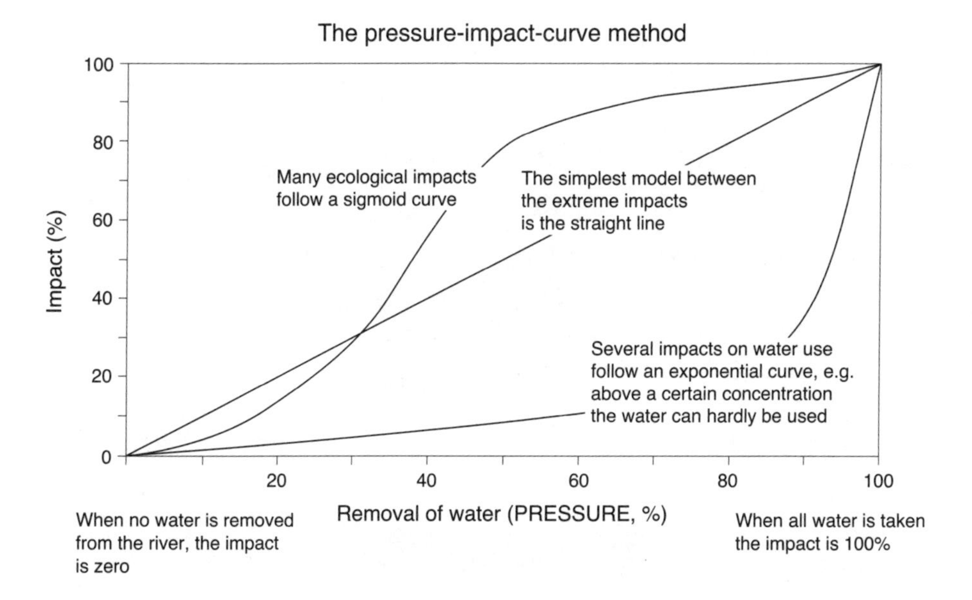

Figure 7.3 The principles of a pressure-impact curve

- Step 5: compute the scores/ranking of the optimal water-level alternatives.
- Step 6: conduct sensitivity analysis of plausible changes in expert-defined value functions and stakeholder-defined weights.

Once the MCA analysis has been completed with the DEFINITE software, the results of the ranking of river flow levels (for a given critical river stretch and period) can be compared to the equivalent levels of hydropower generation (see Figure 7.6). This trade-off curve summarizes the main user conflict between hydropower and other multiple uses. It can serve as a basis to evaluate, for example, the potential for economic compensation.

This can also be used in weighting the different river values, including sensitivity analysis of different individual weights. The results can be used to construct the best possible water-level resultant curve, which ensures that changes to ecosystem values and user interest values are within acceptable limits. The analysis of the pressure-impact curves can be used to adjust the optimum water-level curve, which again will give information on how to adjust the water release pattern from the reservoirs.

The method has been tested as an exercise in different parts of the Glomma River, as well as in different part of the Sesan River, both in Vietnam and Cambodia. The work associated with this method is presented in two technical briefs from the STRIVER project (Barton and Berge, 2008; Nhung et al, 2008) and we recommend these briefs to those readers who are interested in the

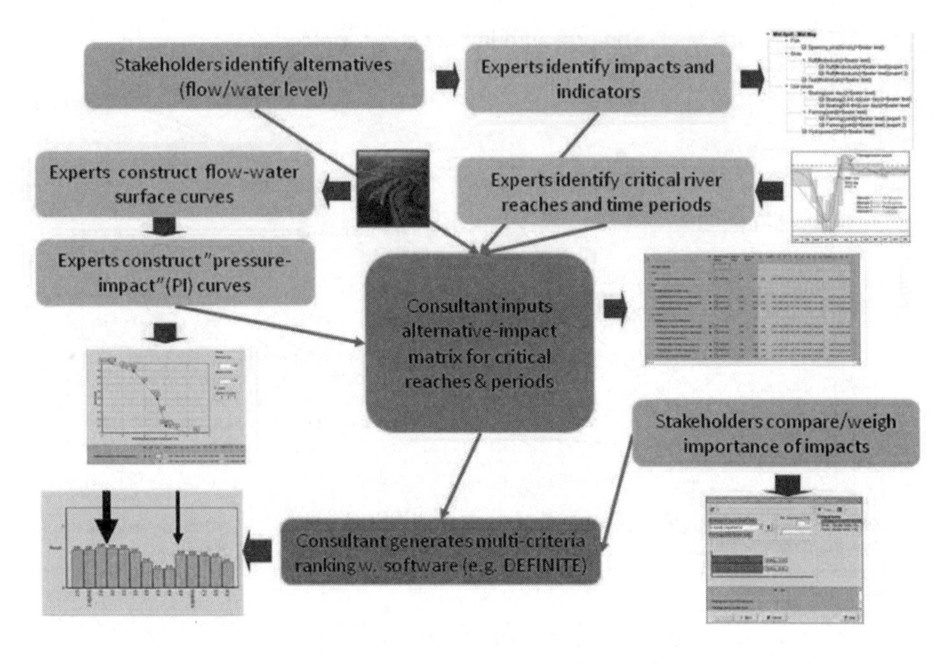

Figure 7.4 Steps related to ranking river flow alternatives by use of DEFINITE MCA software

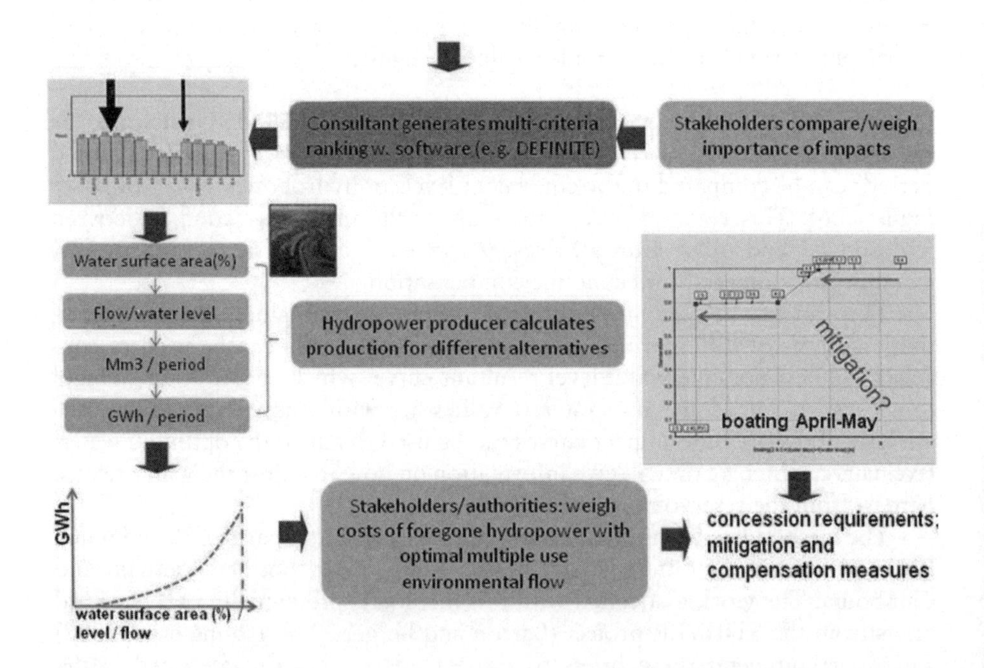

Figure 7.5 Steps related to ranking river flow alternatives by use of DEFINITE MCA software, continued

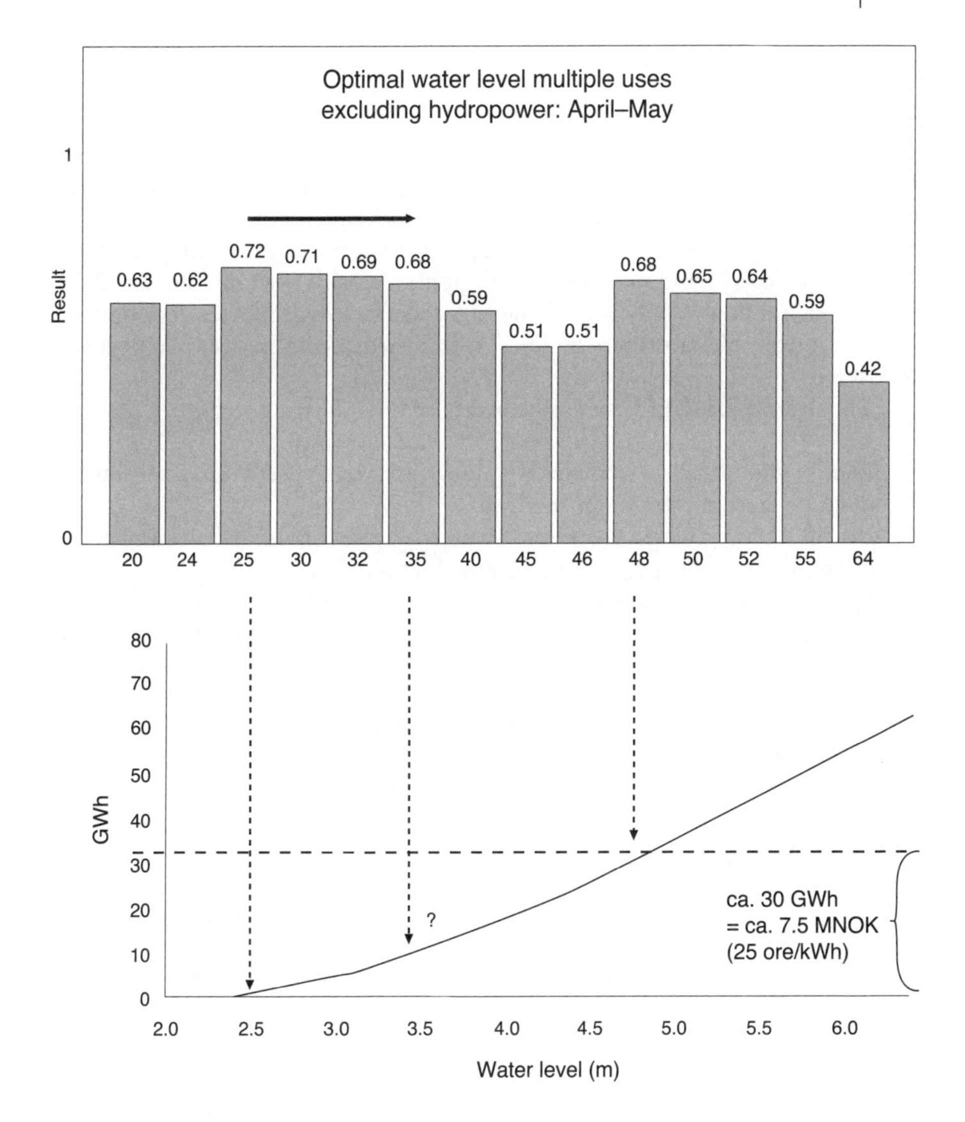

Figure 7.6 The results of the ranking of river flow levels (for a given critical river stretch and period) can be compared to the equivalent levels of hydropower generation

technical details. We therefore do not go into the technical details of the method here, and only give a brief résumé of the test exercises.

The aim of the method is to address the following three items:

1 Identify the ecological values in the river (achieve good ecological potential, which is the environmental requirement in regulated rivers after the European Water Framework Directive).
2 Take into account water use interests (try to fulfil the needs for water level and water flows for a wide spectre of water use).

3 Avoid creating excessive disadvantages for water users through regulation purposes.

The stakeholders involved at different stages should be a combination of local river users and professional experts who can evaluate the impacts of different flow levels (a technical and local knowledge task), and regional and local water authorities and interest groups, who can evaluate the relative importance of different ecological and user interests (a political task). It should be noted that environmental flow is only part of the EIA tasks in a hydropower development, as are, for example, transmission lines, switchyards, dams, power plants, access roads, etc.

The overall PIMCEFA method consists of the following steps:

- Identify the key river ecological values (fish, river bed fauna, periphyton, aquatic macrophytes, water fowl, etc.).
- Identify the key water use values (water supply, fishing, irrigation, bathing and washing, hydropower, flood control, etc.).
- Appoint an expert panel consisting of professional experts and local experienced water users within the fields of river values above, including relevant representatives from water authorities (local, regional and central) (i.e. relevant scientists and stakeholders).
- Draw preliminary optimum water-level curves over the annual cycles that represent the river value you are treating.
- Identify critical periods (i.e. periods when you are confident that the water level needs to be at certain levels): migration periods, spawning periods, sailing depth during boating season, etc.
- For the critical periods, draw pressure-impact curves (i.e. assess maximum damage and minimum damage and draw the most likely curve between these two points).
- Load the pressure-impact curves into a multi-criteria analysis (MCA) tool (as discussed previously, we chose to use DEFINITE; but others also exist).
- Use the MCA tool to evaluate the impact curves of the different river values against each other.
- After trade-offs between the different values are completed, construct the resultant optimum water-level curve.
- Use hydrological models to convert water levels to water flow and provide advice to the hydropower companies on how to plan the dam release.

In order to provide accurate results, the method requires that baseline studies covering the relevant river values, as well as EIA studies are performed. It is important to know how the power plants are planning to operate. However, it is also possible to use the method with fewer available data; but the results may then be more ambiguous. The method can also be used without multi-criteria analysis software, but it will be more laborious to perform the trade-offs. The result could also be more easily influenced by the strongest debater.

The use of scenarios in environmental flow settings

The process of setting environmental flows (or water levels) is an iterative process, where different water flow and water-level regimes (scenarios) are tested against the different ecological values and water use values until the best balanced alternative is achieved. This resultant alternative is, in fact, the environmental flow. Thus, water release scenarios are always used in environmental flow assessment.

A very important parameter in environmental flow assessment is the area of dry river bed at different water flows/water levels. This can be modelled if a network of river profiles or air photos of the river during different water flows are available. If such data are available the river bottom areas that will be dry due to different scenarios, based on released water flow, can be illustrated efficiently.

Other types of scenarios that are relevant to environmental flow assessment are based on the master plan for hydropower development in the different rivers (i.e. how many of the hydroelectric power (HEP) plants in the master plan are likely to be built). In the Sesan River, we took two scenarios into account. The first included all the planned hydropower projects (HPPs) to be built in Vietnam, but not those in Cambodia, while the most likely scenario is that Cambodia will also develop HEP on the Sesan. This is due to a very recent change in the Cambodian authorities' attitude towards hydropower. From having been critical of the problems caused in Cambodia by the Vietnamese HEP regulations, they now have changed their approach by asking the Vietnamese to assist them in forming HEP regulations within the Cambodian part of the Sesan River.

In the latter, the river will be changed into a cascade of large lakes. The environmental flow will then be used to plan the water levels in the lake in the best possible way also in order to produce fish for local livelihoods. In the first scenario, the environmental flow will be adjusted as close as possible to natural flows. Here the re-regulation reservoir downstream of the lowermost HPP will be essential.

SPSI in environmental flow assessment

In the environmental flow setting, it is very important to have wide participation and an open, transparent and replicable process. The holistic types of assessment methods require interaction between all of these groups. In practical hydropower development assessment cases, we advocate the involvement of scientists and the usage of scientific methods.

While the scientists initiated the PIMCEFA method, it has been decisively important to have broad participation from policy-makers and different stakeholders. They steered us towards what was useful and practical, what was possible, what was important and not important, as well as providing alternative ideas which the scientists had not considered. Without their participation, PIMCEFA would have been a method for academics and would never have been used in practical water management.

Wider participation was not only necessary for the development of a useful method, but also proved to be an efficient way of making the method known among water managers. We have observed that the hydropower-related water management authorities in Norway, as well as the larger hydropower companies, know about the PIMCEFA method for assessing environmental flow. Despite the fact that the method is not being fully developed yet, we have already been invited to assist in assessing environmental flow in a new hydropower development project: the Lower Otta HPP. We have already contributed to the terms of reference of the different impact studies under the EIA in order to ensure that the data can be easily integrated within the PIMCEFA method. This example shows that a crucial element for developing a practical management tool is the involvement of key stakeholders.

In the expert panels used for testing the method in the Glomma River (Øyeren and Høyegga), we included representatives from local river users, professional experts, local water management authorities (municipalities, the hydropower association GLB, the county governor) and central water authorities (the Norwegian Water Resources and Energy Directorate). The combination of local and expert knowledge was particularly important in assessing critical periods when certain threshold water levels have to be exceeded. Hydropower experts and water managers provided the frames within which realistic water releases are determined, without excessive negative effects on HPP production. During the course of the project, between 10 and 12 individuals took part in the expert panels in Glomma.

In Cambodia, the expert panel meeting was run with experts from the STRIVER project and with local river users such as fishermen and farmers, as well as some NGOs. These local river users had a clear opinion of what water flow they needed in order to protect their interests, as well as having a clear perception of what was wrong with the regulated conditions, but they could not define critical water levels for river ecology or water use. Within the Kon Tum area of the Vietnamese Sesan, the expert panel consisted of experts from Hanoi and local river users, with much weaker participation from local authorities and hydropower authorities than was seen in Norway. One difficulty was the problem of engaging busy people in a hypothetical test case in remote areas, such as the one upon which the project was based. It would have been much easier to appoint more participants to the panel if it had been a 'real case'.

Lessons learned and practical recommendations

The main lesson learned during this project was that it is very crucial to bring local river users, professional experts and water management authorities into the IWRM project. If not, then the IWRM can easily become just a theoretical academic exercise, providing limited relevance to the practical water management that takes place in all watercourses. The biggest loss in this respect was that we were not in a position to engage the Ministry of Water Resources and

Management (MOWRAM), the government authority responsible for water in Cambodia. Thus, our environmental flow research there became difficult due to a lack of participation from the relevant water management authorities.

Another lesson was that it is not easy to get the necessary and whole-hearted participation of local and regional stakeholders (water authorities, water users, professional experts, etc.) to work within an expert panel in a research project, where everything is a hypothetical case and the work is primarily an exercise. It is much easier to mobilize these types of groups in a real case project as, for example, when elaborating upon a water management plan for a certain river. If, for instance, this project was initiated by the Cambodian and Vietnamese authorities to try and set the environmental flow in the Sesan River, which the hydropower plants then had to follow, it would not have been a problem attaining participation from all levels of stakeholders. In research projects, it is possible that participation may be limited and may only include laymen living along the river and some NGOs, while the water management authorities (local, regional and central) are not properly included. The consequence of such limited participation is that the evaluation of pressure-impact curves (impact assessment) may be carried out; but the assessment of the relative importance of different impacts by relevant stakeholders is incomplete. Planning hydropower development is aimed at providing the best solution for a country as a whole. In the case of the Sesan, however, the input was very one sided, with the majority of the input arising from the poor people living along the river.

In Glomma we included policy-makers from relevant levels, although not everyone who should have been included could participate due to lack of funding for their participation (other than travel and accommodation expenses). Their input was therefore limited to participating in expert panel meetings. A practical recommendation for this type of research project is to put much more emphasis on including the water management authorities (policy-makers) in the project and to allocate money in the project budget for their participation. Clearly, they have substantial practical experience in IWRM that can be highly relevant to the project. However, these authorities are often too busy to allocate their working time to hypothetical exercises without remuneration.

Another way of attaining better and broader-based participation is to link the research project directly to an existing real case management project (i.e. to a hydropower development that is in the development phase). That is the next step which has been taken to further the development of the promising PIMCEFA method. We have linked our work directly to the Lower Otta Hydropower Development Plan (still the Glomma River Basin), which is now at the start-up stage of the impact assessment studies. If this HPP development had started two years earlier, then we could have included it in the STRIVER project and it would have been much easier to secure participation from all relevant stakeholders, as well as policy-makers.

Pros and cons regarding the PIMCEFA method as a tool for assessing environmental flow

The pros of the PIMCEFA method include the following:

- It identifies the most important ecological values and user values.
- It includes the participation of all relevant stakeholders.
- It is a clearly defined version of the expert panel method, which is transparent and replicable.
- It has a computerized multi-criteria analytical tool that makes it easy to compare and co-weight the different river values/impacts.
- It can be used both in water-level settings in reservoirs and as water flow settings in pure river stretches.

The cons of the PIMCEFA method are:

- The optimum water-level curve and the critical periods are not easy to decide upon for all river values, often due to lack of pre-studies, EIAs or skilled personnel.
- The connection between wetted perimeter and flows is rarely established in advance.
- The multi-criteria analysis and co-weighting process of the different water values needs further testing to reveal the method's full potential. This should be done in a 'real case'.

Practical recommendations for the Glomma and Sesan with respect to environmental flow

In the Glomma River, all river stretches (rivers and reservoirs) should be evaluated with respect to environmental flows as part of the river basin management plan (RBMP) in connection with the implementation of the EU Water Framework Directive (WFD). The evaluations should not only be conducted regulation by regulation, but for the total regulation scheme in the river system. This could reveal information that may be useful in future renewals of concessions. Environmental flow should be seen in close connection to other types of mitigation measures. Experienced professional experts and experienced local river users, as well as relevant authorities, should be involved in environmental flow assessment.

The regulations in the Sesan River are more comprehensive than in the Glomma River, whereas the levels of mitigation measures and compensation measures are less developed. Vietnam and Cambodia should develop a joint water management plan for the Sesan River, and environmental flow assessment should be part of this. A joint update of the master plan for hydropower development has been developed, which could serve as a starting point. The work should also be coordinated with the ongoing research of the Mekong River Commission (MRC) and Asian Development Bank (ADB) 3-S Rivers Basin Development

Programme. All river stretches (rivers and reservoirs) should be evaluated with respect to environmental flows as part of the RBMP. It is important to face the fact that it is the reservoirs that will be the most important water bodies in the Sesan River when the total regulations scheme is carried out. If these water bodies can be managed so that water-level fluctuations of less than 3 to 5m are achieved, they can produce a large amount of fish. Stocking programmes may be necessary to compensate for lost spawning conditions and there would be substantial reductions in biodiversity in the main stream river, with many important species disappearing. This is unavoidable with such a comprehensive regulation scheme. In order to abate some of this loss, fish bypass systems could be installed at the dams, but only a few species would be able to use these. The environmental flow in the river stretches should, first of all, be assessed to ensure that the river could function as a spawning and nursery area for reservoir fish, in addition to serving local river uses. Environmental flow could be released via fish ladders to achieve maximum benefit.

In conclusion, environmental flow should be viewed as closely connected with other types of mitigation and compensation measures, with the aim of achieving maximum preservation of local livelihoods and local environment within what is feasible for the regulation purpose and what is accepted by both countries.

References

Barton, D. N. and Berge, D. (2008) *Pressure-Impact Multi-Criteria Environmental Flow Analysis in the Glomma River*, STRIVER Technical Brief No 6, http://kvina.niva.no/striver/Portals/0/documents, last accessed March 2010

www.striver.no/diss_res/files/STRIVER_TB6.pdf

Beinat, E. (1997) *Value Functions for Environmental Management*, Kluwer Academic Publishers, Dordrecht, The Netherlands

Berge, D., Bjørndalen, K., Brabrand, Å., Andersen, R., Dale, S., Bogen. J., Bønsnes, T. E., Martinsen, T., Elster, M., Rørslett, B., Halvorsen, G. and Sloreid, S. E. (2002) *Environmental Studies in Lake Øyeren 1994–2000*, Main Report (in Norwegian), Akershus Fylkeskommune, Oslo, Norway

Berge, D., Barton, D. N., Dang, K. N. and Nesheim, I. (2008a) *Environmental Flows (EF) in IWRM – with Reference to the Hydropower Regulated Glomma River in Norway and Sesan River in Vietnam/Cambodia*, STRIVER Policy Brief No 9, www.striver.no, last accessed March 2010

Berge, D., Barton, D. N., Dang, K. N., Hoang, Phi Thi Thu and Nesheim, I. (2008b) *Environmental Flows (EF) in IWRM – with Reference to the Hydropower Regulated Glomma River in Norway and Sesan River in Vietnam/Cambodia*, STRIVER Technical Report WP8, NIVA Report Lnr. 5693-2008. www.striver.no, last accessed March 2010

Dunbar, M. J., Gustard, A., Acreman, M. C. and Elliott, C. R. N. (1998) *Overseas Approaches to Setting River Flow Objectives*, Environment Agency R&D Technical Report W6-161, Institute of Hydrology, Wallingford, UK

Halleraker, J. H. and Harby, A. (2006) *International Methods for Deciding Environmental Flow – Which of These Are Applicable in Norway?*, NVE Miljøbasert vannføring, Report 9 2006, Norway

Janssen, R. and Herwijnen, M. V. (2007) *DEFINITE 3.1: A System to Support Decisions on a Finite Set of Alternatives (Software Package and User Manual)*, Institute for Environmental Studies (IVM), Vrije Universiteit, Amsterdam

Jowett, I. G. (1997) 'Instream flow methods: A comparison of approaches', *Regulated Rivers: Research and Management*, vol 13, pp115–127

King, J. M. and Louw, D. (1998) 'Instream flow assessment in South Africa using the building block methodology', *Aquatic Ecosystems Health and Management*, vol 1, pp109–124

King, J., Tharme, R. E. and Brown, C. (1999) *Definition and Implementation of Instream Flows*, World Commission on Dams, Thematic Report, Southern Waters Ecological Research and Consulting, Cape Town, South Africa, p63

Nhung, D. K. and Hoang, P. T. T. (2008) *Pressure Impact Multi-Criteria Environmental Flow Analysis in the Sesan River*, STRIVER Technical Brief No 7, http://kvina.niva.no/striver/Portals/0/documents/STRIVER_TB7_EFSesan.pdf, last accessed March 2010

Scruton, D. A., Pennell, C. J., Robertson, M. J., Ollerheard, L. N. M., Clarke, K. D., Alfredsen, K., Harby, A. and Mckinley, S. R. (2005) 'Seasonal response of juvenile Atlantic salmon to experimental hydropeaking power generation in Newfoundland, Canada', *North American Journal Fish Management*, vol 25, pp964–974

Tharme, R. E. (2003) 'A global perspective on environmental flow assessment: Emerging trends in the development and application of environmental flow methodologies for rivers', *River Research and Applications*, vol 19, pp397–441

8

The Science–Policy–Stakeholder Interface and Transboundary Water Regimes

Geoffrey D. Gooch and Alistair Rieu-Clarke

Introduction

Two of the case basins studied in this book are international (transboundary) rivers: these are the Sesan River in South-East Asia, which flows from the Central Highlands of Vietnam into north-east Cambodia, and the Tagus River, which flows from Spain into Portugal. As was also noted in Chapter 6, while the Tungabhadra River in India is not an international river, but an interstate river, it also demonstrates some of the characteristics of a transboundary river.

Transboundary rivers are a special challenge for water management as they involve cooperation between two different sovereign states, each with their own legal systems and institutions. In some cases, such as the Sesan River, they may also straddle two different political systems. This often complicates the cooperation necessary for efficient water management, and it does, of course, also place special demands on the science–policy–stakeholder interface (SPSI). In this chapter we will first discuss some generic problems of transboundary rivers and then continue with more detailed descriptions of our two case rivers, the Sesan and Tagus. We will show how we went about identifying and studying the special challenges of these rivers, and how we attempted to actively help formulate recommendations for water management in these rivers. In line with the theme of this book, we also analyse the SPSI in these rivers and discuss ways of improving it. The work reported in this chapter has been conducted in close cooperation with that in Chapter 3 (participation) and Chapter 4 (scenarios), and we refer back to those chapters so that the reader can, when necessary, locate more details of the methodologies used for participation and scenario-building.

Transboundary rivers

There are 263 international river basins that cover more than 45 per cent of the Earth's land surface (Giordano and Wolf, 2002), with over half of the world's population living in these transboundary river basin areas. Because of their importance for the people residing in the countries that they flow through, as well as for biodiversity and environmental resources, the successful management of these international water resources is a major challenge (Olem and Duda, 1995). The management of transboundary waters creates specific problems over and above the general challenges of water management, as these rivers, lakes and other water bodies (such as underground aquifers) involve the interaction between at least two different legal, political and administrative systems. These transboundary waters differ widely in the challenges that they pose: in some parts of the world, water quality is the major problem; in others there is sufficient water, but water quality or the effects of water use by different, often competing sectors is a problem. Therefore, while a considerable amount of work has been conducted in transboundary waters and rivers during the last decades, much of it has focused on potential conflicts over water quantity, predominantly in the Middle East, Africa, India and America, where lack of water is the main driver. In the cases that we focus on in this chapter, the Tagus certainly fits into this category and, as has already been shown in Chapter 6, there is a general shortage of water for all uses in this basin. In the Sesan, however, the main problem is not a lack of water, but the uneven distribution of water released from hydroelectric power (HEP) dams in the upriver country of Vietnam.

Law and transboundary water regimes

Historically, two alternative claims of state entitlement over the use of transboundary waters existed: absolute territorial sovereignty and absolute territorial integrity (Berber, 1959). Absolute territorial sovereignty was based on the principal that a state could use all water within its boundaries as it pleases. This is obviously a point of view often taken by upstream countries, which may also claim that they cannot be held responsible for damage resulting from their actions on down stream states. Such a viewpoint was known as the Harmon Doctrine, named after an advisory opinion by US Attorney-General Harmon in 1895 (Wouters, 1997). The alternative view – 'absolute territorial integrity' – is that the upstream state cannot develop transboundary waters if it will cause harm to the downstream state (Utton, 1973). Neither of the two alternative claims has received much support for a number of reasons (Rieu-Clarke, 2005). Watercourse states are not easily segregated into upstream and downstream. A state may be both upstream and downstream on the same river, or may be both upstream and downstream on a number of rivers flowing through the territory of that state. In addition, practice shows that claims of absolute territorial sovereignty and absolute territorial integrity are used as bargaining positions by states, prior to reaching a solution based on compromise (McCaffrey, 1996).

The doctrine of limited territorial sovereignty, a compromise solution, is now widely accepted by states as the basis for determining state entitlement over the use of transboundary waters (Tanzi and Arcari, 2001). As articulated in the 1992 Rio Declaration, the principle provides that:

> States have, in accordance with the Charter of the United Nations and the principles of international law, the sovereign right to exploit their own resources pursuant to their own environmental and developmental policies, and the responsibility to ensure that activities within their jurisdiction or control do not cause damage to the environment of other states or of areas beyond the limits of national jurisdiction.

The concept of limited territory provides the basis for the cornerstone principle of international water law, which obliges states to utilize their transboundary waters in an *equitable* and *reasonable* manner. As we will see in this chapter, the question of the right to exploit their own (water) resources for HEP production is especially stressed by Vietnam in the case of the Sesan River, while the downstream country, Cambodia, focuses on the second right: that a state should not cause damage to other states. How, then, to manage transboundary rivers and international waters when there are competing interpretations of the rights and responsibilities of states, and at the same time a lack of a single legal and political authority that can decide on these issues?

One of the responses to transboundary water management is the development of international regimes (Levy et al, 1995). Regimes can enable states to manage policy in a specific issue area in conditions where a central authority is lacking, such as in an international context in which every state is sovereign. Krasner (1983, p2) states that international regimes are:

> ... implicit or explicit principles, norms, rules and decision-making procedures around which actors' expectations converge in a given area of international relations. Principles are beliefs of fact, causation and rectitude. Norms are standards of behaviour defined in terms of rights and obligations. Rules are specific prescriptions or proscriptions for actions. Decision-making procedures are prevailing practices for making and implementing collective choice.

Here we can see that the rights and obligations of the two legal interpretations mentioned above are also seen as part of a regime. While Krasner's definition of a regime contains many of the aspects still considered relevant, such as norms, rules, principles and decision-making procedures, his definition has been criticized as being too vague (Levy et al, 1995) and others have suggested instead that 'Regimes are institutions with explicit rules, agreed upon by governments that pertain to particular sets of issues in international relations' (Hasenclever et al, 1997, p12).

This is, of course, a more limited definition of regimes as it only covers those formal agreements made by governments, and we would argue that a regime can also be construed by non-governmental actors too. The concepts of norms and rules are central to international regimes, and Porter and Brown (1996, p16) define a regime as 'a system of norms and rules that are specified by multilateral agreement among relevant states to regulate national action on a specific issue or set of interrelated issues'.

We can therefore choose to focus on these systems of norms and rules and/or on formal regimes as contained by agreements – for example, treaties, conventions or protocols that are defined under the 1969 Vienna Convention on the Law of Treaties as being 'an international agreement concluded between states in written form and governed by international law, whether embodied in a single instrument or in two or more related instruments and whatever its particular designation'.

Once in force such agreements are binding on the parties to it, and can be said to constitute a formal regime, and the underlying provisions must be carried out in good faith. At a global level, the 1997 Convention on the Non-Navigational Uses of International Watercourses represents a codification and progressive development of rules and principles for enabling and sustaining transboundary cooperation. To date, the convention counts 17 contracting states – 19 short of the number required for entry into force. In addition, there are other multilateral environmental agreements at the global level that partially relate to transboundary waters (e.g. the Biodiversity Convention, the Climate Change Convention and the Ramsar Wetlands Convention). At a regional level, watercourse conventions such as the United Nations Economic Commission for Europe (UNECE) Helsinki Convention, the European Community Water Framework Directive (WFD) and the Southern African Development Community (SADC) Protocol have been adopted to strengthen the implementation of watercourse agreements. These regional instruments provide more detailed provisions, particularly in relation to implementation instruments such as monitoring, assessments, public participation and the establishment of basin-specific arrangements. In addition to the above-mentioned laws, 158 of the 263 international river basins are covered, at least in part, by basin or sub-basin agreements, although few such agreements provide sufficient provisions to reconcile competing interests amongst states (Giordano and Wolf, 2002). As noted above, regimes may also be founded on non-binding instruments that seek to promote shared understanding. Such instruments include United Nations General Assembly Resolutions and outcomes of international conferences (e.g. the 1972 Stockholm Declaration and the 1992 Rio Declaration). A combination of binding and non-binding instruments may also shape a regime.

In the study of regimes in a broader sense than the formal definition of Hasenclever et al (1997), three distinct positions have emerged (Gooch et al, 2002), arguing that regimes are best studied from a behavioural, formal or cognitive perspective. From the behavioural perspective, the empirical evidence of a regime is to be found in its ability to shape the behaviour of actors: if it does not shape the actors' behaviour, it is not a regime. While this may seem to imply a formal agreement or influence, this is not necessarily so. Informal norms and values, if shared, can also influence behaviour. The second position focuses on the legal side of a regime: the explicit rules agreed upon by actors relating to a specific issue area. The legal agreements given as examples above represent this form of regime. The third perspective focuses on shared understandings (Hasenclever et al, 1997), and here we can see that the SPSI, which focuses on the exchange of ideas and knowledge between actors in water management, can be used as an example for a cognitive regime. In our studies of regimes on the Sesan and Tagus rivers, we utilized aspects of all three perspectives, as this approach helped us to understand how and why transboundary cooperation in water management does, or does not, occur in the case basins.

Actor networks in transboundary regimes

If, as the definitions imply, we should see regimes as agreements, formal or informal, between actors (states or other), then our first step in an analysis of transboundary water regimes must be to identify these actors. This can be done in a number of ways. We could, following the formal definition, simply see who had signed the agreements that constitute the regime and then count them as actors. This would probably not give a complete picture of all the actors, however: we would need to know who signed the agreement and their position in the government or department. We would also need to know something about why they signed, as this could probably help us to understand the chances of successful implementation. If the agreement was signed and the regime created as a form of symbolic politics (i.e. without any clear ambition of fulfilling obligations but only as a show), then the chances of successful implementation will be slight. Another way would be to follow negotiations on an agreement and takes notes on who participated and how. This anthropological approach has much to be said for it, but it is also time consuming and in the case of our transboundary basins was not possible due to time and resource restraints, as well as the sensitive political situation in one of our cases: the Sesan River.

In order to focus on the actors in the transboundary regimes and their interactions, we therefore used actor network theory (ANT) as a guiding tool in our efforts to identify the actors. This allowed us to begin with a rough idea of who and what we believed were part of the regime, and then to move on through interviews and fieldwork to a greater understanding of the regimes. ANT also helped us to look at the situation beyond the human actors and to take into account other non-human actors. This may sound, at first, somewhat strange, particularly for those who have learned that an actor is someone with a will, a cognizant being. Moreover, this becomes even more complicated when the non-human actor is not another living entity. Instead, the approach leads us to focus not only on the human actors, but also on non-human entities such as HEP dams, legal systems, riversides, agriculture techniques, communication tools and, naturally, the rivers themselves. Without going into the details of the approach, we can say that the method involves identifying what is seen as the main influence, whether human or non-human, and then following up this actor's dealings with others (human and non-human) within the network. As we have been using heavily modified rivers as our case basins (i.e. rivers in which hydroelectric power stations have been constructed or dams for irrigation), it seemed to us natural to begin with these structures and to follow the networks that led from and to them. In the case of the Sesan, for example, our focus on the HEP dams was the result of preliminary desktop studies, as well as field trips conducted during the early stages of the project. This led us to concentrate on the networks around the dams, which included local villages, government departments, non-governmental organizations (NGOs), legal documents, environmental flow levels (see Chapter 7), information about the dams, dam engineers, etc. This put the HEP dams at the centre of our network. Other alternatives could have been to begin with the local villages, which would probably have led us to look at other actors such as roads and infrastructure, departments dealing with relocation policies, housing techniques and conditions, and so on.

As already mentioned, ANT aided us in identifying these actors. ANT has generated a certain interest as a possible means of integrating the 'natural' and the 'social' in one and the same analytical framework. The potential advantages of this combination of theoretical approaches for governance and transboundary water regimes is that ANT seems to provide a means of integrating the perspectives of the 'natural' and 'social' sciences, a goal that has been much discussed in water management but seldom practically achieved (see Chapter 2). One of the central tenets of ANT is that of 'general symmetry and symmetrical analysis', which refers to the need for non-human elements (of a network) to be treated analytically in the same way as the social and human elements (Law, 1992). This claim has resulted in a lively discussion and certain consternation (e.g. Vandenberghe, 2002; McLean and Hassard, 2004), and a number of authors have pointed out that human actors perceive and act upon their world in ways that non-humans perhaps cannot (Bruun and Langlais, 2003). However, it seems clear to some of us working in water management that a conglomeration of actors is precisely what we have to deal with (Gooch, 2004). We need to overcome the barriers erected by the different academic disciplines working within water governance and to combine the insights of all of them into an approach that utilizes complementary aspects of natural and social sciences (Gooch and Stålnacke, 2006). This represents an important challenge for improving the SPSI, as noted in Chapter 2. Studies of international and global water systems need to take into account a wide variety of actors and sources, from government agencies, NGOs, biologists, research reports, dams, studies of water quality and quantity, and many more. This, of course, raises a difficult question. Can technologies such as hydroelectric power stations, desalination plants, irrigation canals and wastewater treatment plants be considered political phenomena in their own right (Pels et al, 2002), and can there be a 'politics of things' (Winner, 1986)? These questions cannot be answered outright as they need to be the focus of empirical studies. We need, however, to keep an open mind about the type of actors who are influential, and to be prepared to accept that dams and statistics can play major roles, just as governments and environmental administrators can.

ANT has been used in organizational analysis, business studies, informatics, health studies, geography, sociology, anthropology and economics, and, during recent years, in some studies of the environment. The studies produced using ANT differ considerably in their approach, some focusing on one aspect of the theory, others on another aspect. There is hardly one single method, but rather a variety of related approaches. ANT is sometimes called the 'sociology of translation' (Callon, 1986), yet might well be called the 'politics of translation' for, as Law (1992) notes, 'actor-network theory is all about power – power as a (concealed or misrepresented) effect, rather than power as a set of causes' (Law, 1992), and politics has always been, and still is, all about power. Water governance is also about power, at least from the perspective of law, politics and policy. Yet, actors gain power through their relations with others, and in many cases 'power is not in the hands of one person who can exercise it alone and totally over others' (Foucault, 1980), but is a result of those relationships. This should lead us to look closely at the relationships between actors, and between actors and other entities. Starting from the HEP dams in Vietnam and from the Tagus River itself in Spain and Portugal, we have tried to follow each actor's paths of influence. Figure 8.1 shows the results of the analysis for the Sesan River in Vietnam, where the HEP stations are in the lower left-hand part of the figure, while the government of the country is in the upper right-hand section.

Communication in transboundary regimes

Communication is a central aspect, both for the regimes and within the SPSI. In the case of water management, and specifically in transboundary water management contexts, this communication becomes even more central as it must provide a means of interaction between the large number of actors coming from different scientific, managerial, legal, political, administrative and other systems. The potential for conflict based on inefficient communication and resulting misunderstanding is high in this context. In the United Nations Economic and Social Commission for Asia and the Pacific (UNESCAP) report, Guidelines on Participatory Planning and Management for Flood Mitigation and Preparedness (UNESCAP, 2003), five cases of potential water conflict were identified (potential or resultant water conflicts are also discussed in Chapter 3). These were:

1 relationship conflicts;
2 data conflicts;
3 value conflicts;
4 structural conflicts;
5 interest conflicts.

All of these potential conflict factors are strongly dependent upon and influenced by the level of efficiency of communication systems. For example, relationship

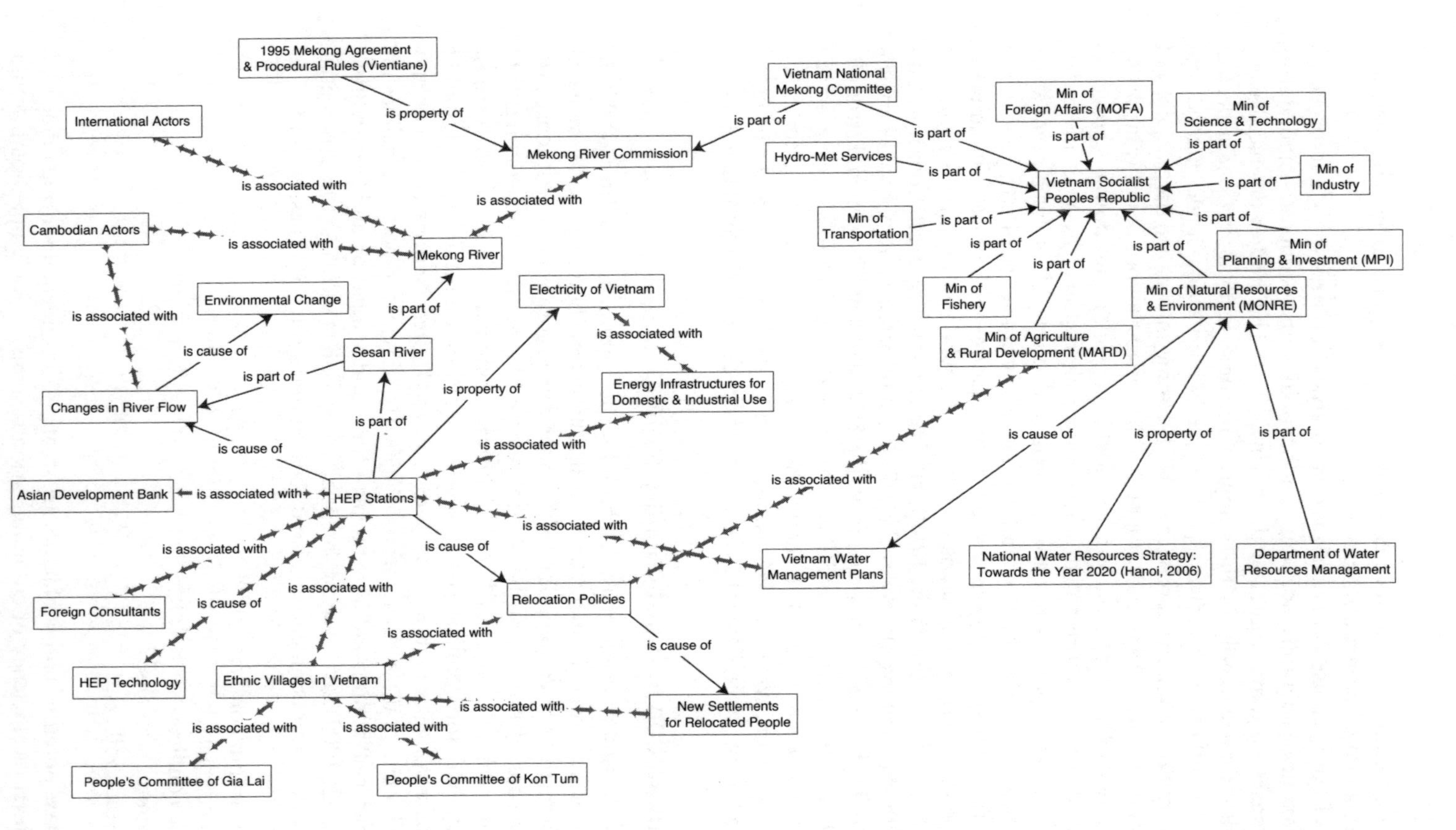

Figure 8.1 Actor networks on the Sesan River in Vietnam

Source: G. D. Gooch

conflicts (point 1) can be made worse if there is a lack of understanding caused by poor communication. In the same way, data conflicts (point 2) can result in a lack of common information, or unreliable information and value conflicts (point 3) may, while they can hardly be solved through better communication, be diminished if a basic understanding of actors' beliefs and values is achieved, as can interest conflicts such as competition for water allocation (point 5). These potential areas for conflict are not only influenced by communication processes, but also influence these processes. Value and interest conflicts strongly influence the formation, communication and reception of information, as will be discussed in more detail below.

Information is vital for transboundary regimes and SPSI; without access to information, decisions taken will be haphazard, unreliable and perhaps inaccurate. Information, however, never seems to come in the right quantity. Sometimes there is too little information available; sometimes the amount of information is so vast that it makes it difficult to see what is, or is not, relevant. In addition, it is often difficult to see how to identify reliable and relevant knowledge from the general flow of information. This information is rarely objective, as much of it tends to serve specific interests and should be seen as a form of power, a way of influencing others in their behaviour and thoughts. As information constitutes a form of power, access to information can be limited or encouraged by those who manage it. Information can also be consciously distorted to serve the interests of one or other group or individual, as can the processes of communication. Within SPSI, a third major problem is that the receiver of information may not be able or willing to understand it, or may not be prepared to accept it. Gooch et al (2002) note that transboundary water management has a number of specific characteristics concerned with information. These include:

- the central role of knowledge about environmental conditions;
- the necessity for cooperation and communication in problem solving;
- the need for actors to share information and harmonize databases;
- in cases where there are joint transboundary river commissions, the importance of collecting and communicating scientific information.

Within transboundary regimes and the SPSI, we claim that information consists not only of facts and figures, but also of beliefs, values, opinions and organizational procedures. We also claim that we need to distinguish in the SPSI between information *presentation*, which is a one-way process through which one party in transboundary water management provides data to another, and information *communication*, which is a two-way transfer of information, opinions, beliefs, feelings, culture, etc. between individuals, groups or societies.

Law in transboundary regimes

Law plays an important role both in regimes and within the SPSI. Additionally, the effectiveness of communication and positive interaction amongst actors will be contingent upon there being effective legal frameworks in place. As a result, there is a close link between actor networks, communication and law within the context of transboundary regimes.

At the broad governance level, law can support the rights of access to justice, information and participation in decision-making, as well as rights that influence the way in which actor networks are established; and communication between science, policy and stakeholders can be secured. At the river basin level, national and international laws set out certain substantive and procedural rules by which competing claims over the uses of water is reconciled. In relation to functional and geographic scope, laws define what is covered by a particular formal agreement (e.g. river basin, sub-basin, mainstream, surface water, groundwater). Substantive principles stipulate standards that determine which use should prevail over other competing claims. Procedural rules provide a means by which different actors can work together to identify and reconcile competing claims. Such rules include conducting environmental impacts assessments, notification of planned measures, and regularly exchanging data and information. Communication is therefore pivotal to ensure the effective implementation of the substantive legal principles.

The focus of the following section is the contribution of law, both at the governance and basin levels, within the Sesan and Tagus rivers.

Law in the Sesan and Tagus rivers

The Sesan River

A central element of achieving good governance and improving the SPSI framework is the need to ensure effective *participation* in decision-making procedures. At the international level, no provision for stakeholder participation is contained within the 1995 Mekong Agreement, although it could be maintained that such participation is essential for the fulfilment of many of the objectives under the instrument. Accordingly, the Mekong River Commission (MRC) has taken significant steps during recent years to enhance stakeholder participation in its various activities. Within Vietnam, there is a range of legislation relating to public participation, including Government Decree 79/2003/ND-CP on Democracy at Grassroots Level, the Law on Environmental Protection, the 2003 Construction Law (which details the collection of public opinion on master plans), and Decree No 88/2003, which provides a clear legal basis for the formation and management of NGOs (UN Country Team, 2004). In Cambodia, according to the government's political platform of 2003 to 2008, 'civil societies shall play the role of efficient partners with the government in building the country'. Additionally, a law on NGOs and associations has been prepared and is expected to be submitted for consideration of the assembly by the fourth legislature (2008 to 2013).

While provisions on civil society access to redress and remedy are not contained in the 1995 Mekong Agreement, both the Vietnamese and Cambodian constitutions recognize that their citizens are equal before the law, and enforceable rights for individuals are in place. In Vietnam, supporting legislation, including Decision 734/TTg of 1997, establishes organizations under the Ministry of Justice that provide *pro bono* legal services to the poor and disadvantaged people, especially women victims of domestic violence, juveniles, ethnic minorities and preferential policy groups (heroic mothers, under 18-year-old children of war martyrs, invalids, etc.). No such provision is included in Cambodian legislation, although the Governance Action Plan II 2005-8 acknowledges that access to legal aid is important, and highlights it as a strategic goal with respect to judicial and land reform.

At the basin level, both Vietnam and Cambodia are party to the 1995 Mekong Agreement, which provides the basic principles by which these states cooperate over their shared waters. Both countries are also party to numerous international and regional conventions related to water resources and the environment. Concerning formal demands on communication and the SPSI, provisions related to notification and consultation are included in the 1995 Mekong Agreement and are part of customary international law. More detailed provisions for the Lower Mekong Basin have been developed through the Procedures for Notification, Prior Consultation and Agreement, which were approved by the MRC Council in 2003. In relation to public access to data and information, there are no explicit provisions within the Mekong Agreement. However, an objective of the Procedures for Data and Information Exchange and Sharing is to 'make available, upon request, basic data and information for public access' (MRC, 1995). In addition, the MRC as custodian of information has established the MRC Information System, designed to support planning, development, decision-making and monitoring activities under the Mekong Agreement.

The Tagus River

At the basin level, Spain and Portugal have long been engaged in treaty development with respect to their transboundary watercourses, dating back to the Limit Treaty of 1864, (concluded by an exchange of notes in 1912). Such developments have culminated in the 1998 Albufeira Agreement, which governs 'the hydrographic basins [including associated groundwater] of the rivers Miño, Limia, Douro, Tagus and Guadiana' (Art. 3(1)). The purpose of the agreement is 'the protection of the surface waters and groundwaters and the aquatic and terrestrial ecosystems that directly depend on them and for the sustainable use of the water resources of the hydrographic basins' (Art. 2(1)). In accordance with its purpose, the Albufeira Agreement recognizes the right of each of the parties to the sustainable use of water resources, as well as their obligation to protect such waters, and to implement measures to prevent, eliminate, mitigate and control transborder impacts (Art. 15(1)). The agreement also obliges the parties to define 'the flow regime necessary to guarantee the satisfactory conditions of the waters' (Art. 16(1)). In addition, the parties must 'adopt,

Figure 8.2 The Tagus River in Portugal

Source: G. D. Gooch

individually or jointly, the technical, legal, administrative and other measures' necessary to, *inter alia*, achieve satisfactory condition of the waters; prevent degradation of the waters and control pollution; prevent, eliminate, mitigate or control transborder impacts; ensure that the use of the water resources are sustainable; and promote rational and economic use. In terms of water quality and pollution, the agreement also obliges the parties to comply with quality targets and standards of waters classified under European Community law. The most recent version of the convention is currently in the ratification process and includes greatly increased detail on the flow regimes applicable to each of the basins that it covers, including daily flow rates where appropriate.

Both countries are bound by the terms of the Water Framework Directive, and from that perspective are obliged to establish basin management administrations and to manage pollution control and abstractions at the basin level. Basin management has existed in Spain since 1926, and this has therefore made it receptive to the transposition of the Water Framework Directive. In Portugal, however, detailed regulations for basin management do not yet exist, resulting in the poor application of the existing law. In Spain, integrated water resource management (IWRM) has been implemented largely through Royal Decree No 1/2001, the Spanish Water Law, which integrates surface and groundwaters and reaffirms the catchment as the basis for water management. Portugal was one of five member states where the European Court of Justice rules against them for the implementation of the Water Framework Directive (case C-118/05; see also case C-251/03 where the European Court of Justice condemned Portugal for not complying with the quality parameters laid down in the Drinking Water Directive).

At the broader governance level, both Spain (29 December 2004) and Portugal (9 June 2003) have ratified the Århus Convention. As was explained in Chapter 3, the convention obliges public authorities, in response to a request for environmental information, 'to make such information available to the public, within the framework of national legislation' without an interest having to be stated (Art. 4(1)). Such information must generally be made available within one month after the request has been submitted. Certain rights of refusal are included within the Århus Convention, but these must be interpreted in a restrictive manner.

As regards the exchange of information between riparian states, the parties to the 1998 Albufeira Convention, through the convention's commission, must regularly and systematically exchange data and information on matters covered by the convention, including the management of the basins' waters; activities that may have a transboundary impact; and legislation, organizational structures and administrative practices. Detailed provisions are contained in Annex 1. The parties are also obligated to provide the European Commission with 'all the information it needs to fulfil its terms of reference and responsibilities', including information on how the activities of the convention are implemented nationally, and those activities that might have a transboundary impact. In addition, the parties are obliged to produce an annual report on relevant activities and an update on the status of national implementation. With respect to extreme events, the Albufeira Convention also obliges Spain and Portugal to 'establish or improve joint or coordinated communication systems to transmit early warning or emergency information, to prevent or correct the situation in question and to take pertinent decisions'. The parties must provide information on the entities and procedures within each of their countries for the transmission of information on early warning and emergency situations. With respect to public rights of access to information under Albufeira, the parties are obligated to 'create the conditions to make available to anyone who submits a reasonable application the requested information on matters covered by' the convention, but may refuse to do so if the following will be affected: national security; the confidentiality of procedures carried out by public authorities; the international relations of the state; the general security of the population; the confidentiality of legal proceedings; or commercial/industrial confidentiality (Art. 6).

Scenarios as a policy tool in transboundary water regimes

In Chapter 4 the use of scenarios in water management was discussed and it was claimed that scenarios present a way for the policy-maker or manager to test ideas about possible futures through exercises that can make clearer the

Figure 8.3 Local villagers discussing transboundary issues on the Sesan River in Cambodia

Source: G. D. Gooch

probable results of certain courses of action. As noted in Chapters 3 and 4, they can also be used as a means of improving public and stakeholder participation. If these groups are involved in the formulation and evaluation of scenarios, they can provide insights not readily available for policy-makers, as well as increase the level of social learning.

In the studies of transboundary regimes in the Sesan and Tagus basins, we used scenarios as a means of two-way communication between scientists (the project team) and stakeholders, NGOs and the public. Since water managers and decision-makers at the local and regional levels were included in the stakeholder workshops, we also interacted with these groups. In this way all three groups of actors in the SPSI were included in these discussions. As described in Chapter 4, the process was based on interactive participatory scenarios and involved a three-step process through which the project team first formulated preliminary scenarios that were then presented at the first stakeholder workshop. These preliminary scenarios were then modified after comments from the groups and updated versions were presented at the second round of stakeholder workshops. These updated versions were then once again modified after comments from the groups and final versions were presented during the last round of stakeholder workshops. The final workshop and discussion of future scenarios in the Sesan took place in Vientiane, Laos, in 2008 at a meeting where representatives from both Vietnam and Cambodia met to discuss a common future for the Sesan. This was in itself a major achievement as relations between the two countries have at times been somewhat strained. What could be observed during the three-year process was that the Mekong River regime, which is based on agreements primarily on the main course of the Mekong between most (though not all) of the riparian states, was complemented by a budding regime around the Sesan itself, a regime not based on formal agreements but on the development of discussions that might, hopefully, in time lead to a common understanding of transboundary water management problems, and with it common values and norms.

Policy recommendations for transboundary regimes and the SPSI

The major challenges to implementing successful transboundary water regimes are connected with problems of implementation of policies, laws and communication strategies. While considerable resources have been focused on formulating policies, etc., less has been invested in implementation strategies and evaluation of policies. Management processes can be improved through:

- An inventory of all actors (human and non-human) and of all institutions (formal and informal). Predefined concepts and opinions should be avoided in this process so as not to miss important aspects.
- Objective evaluations of the effectiveness of legal measures, communication strategies and policies. These evaluations should analyse policy formulation (who has been involved and how), policy processes, policy aims and policy outcomes.

References

Beaumont, P. (2000) 'The 1997 UN Convention on the Law of Non-Navigational Uses of International Watercourses: Its strengths and weaknesses from a water management perspective and the need for new workable guidelines', *Water Resources Development*, vol 16, no 4, pp475–495

Berber, F. J. (1959) *Rivers in International Law*, Stevens & Sons, London

Bruun, H. and R. Langlais (2003) 'On the embodied nature of action', *Acta Sociologica*, 46(1): 31-49

Callon, M. (1986) 'Some elements of a sociology of translation. Domesticating the scallops and fishermen of St Brieuc Bays', in J. Law *Power, Action, Belief: a New Sociology of Knowledge?* Routledge and Kegan Paul, London, pp196–233

Claussen, E. (2001) 'Global environmental governance', *Environment*, vol 43, no 1, pp28–34

Foucault, M. (1980) *Power/Knowledge*, Pearson, Harlow, UK

Garner, R. (2000) *Environmental Politics*, Macmillan Press Ltd, Houndsmill, UK

Giordano, M.A. and Wolf, A.T. (2002) 'The World's Freshwater Agreements: Historical Developments and Future Opportunities', in *Atlas of International Freshwater Agreements*, Nairobi, Kenya

Gooch, G. D. (2004) 'The communication of scientific information in institutional contexts: The specific case of transboundary water management in Europe', in

J. G. Timmerman and S. Langaas (eds) *Environmental Information in European Transboundary Water Management*, IWA Publishing, London

Gooch, G. D., Höglund, P., Roll, G., Lopman, E. and Aliakseyeva, N. (2002) 'Review of existing structures, models, and practices for transboundary water management', Paper presented to the Second International Conference on Sustainable Management of Transboundary Water in Europe, Miedzyzdroje, Poland, 21–24 April 2002

Gooch, G. D. and Stålnacke, P. (eds) (2006) *Integrated Transboundary Water Management in Theory and Practice: Experiences from the New EU Eastern Borders*, IWA Publishing, London

Hasenclever, A., Mayer, P. and Rittberger, V. (1997) *Theories of International Regimes*, Cambridge University Press, Cambridge, UK

Holsti, K. J. (1995) *International Politics: A Framework for Analysis*, Prentice-Hall International, London

Krasner, S. (1983) 'Structural causes and regime consequences: Regimes as intervening variables', in *International Regimes*, Cornell University Press, Ithaca, NY, pp1–22

Law, J. (1992) 'Notes on the theory of the actor network: Ordering, strategy and heterogeneity', *Systems Practice*, vol 5, pp379–393

Levy, M. A., Young, O. R., and Zürn, M. (1995) 'The study of international regimes', *European Journal of International Relations*, vol 1, no 3, pp267–330

McCaffrey, S. C. (1996) 'The Harmon Doctrine one hundred years later: Buried, not praised', *Natural Resources Journal*, vol 36, pp549–590

McLean, C. and Hassard, J. (2004) 'Symmetrical absence/symmetrical absurdity: Critical notes on the production of actor-network accounts', *Journal of Management Studies*, 41, no 3, pp493–519

MRC (Mekong River Commission) (1995) *Agreement on the Cooperation for the Sustainable Development of the Mekong River Basin*, www.mrcmekong.org/agreement_95/agreement_95.htm, last accessed March 2010

MRC (2003) *Procedures for Notification, Prior Consultation and Agreement*, www.mrcmekong.org/download/agreement95/Procedures_Guidlines/Procedures-Notification-Prior-Consultation-Agreement.pdf, last accessed March 2010

Olem, H. and Duda, A. M. (1995) 'International watercourses: The World Bank looks toward a more comprehensive approach to management', *Water Science Technology*, vol 31, no 8, pp345–352

Pels, D., Hetherington, K. and Vandenberghe, F. (2002) 'The status of the object: Performances, mediations, and techniques', *Theory, Culture and Society*, vol 19, no 5/6, pp1–21

Porter, G. and Brown, J. W. (1996) *Global Environmental Politics*, Westview Press Inc, Boulder, CO

Rieu-Clarke, A., (2005) *International Law and Sustainable Development – Lessons from the Law of International Watercourses*, IWA Publishing, London

Tanzi, A. and Arcari, M. (2001) *The United Nations Convention on the Law of International Watercourses – A Framework for Sharing*, Kluwer Law International, London

United Nations Country Team, (2004), United Nations Common Country Assessment for Vietnam, www.undp.org/asia/country_programme/CCA/CCA-Vietnam2004.pdf, last accessed March 2010

UNESCAP (United Nations Economic and Social Commission for Asia and the Pacific) (2003) *Guidelines on Participatory Planning and Management for Flood Mitigation and Preparedness*, New York

Utton, A. E. (1973) 'International water quality law', *Natural Resources Forum*, vol 13, pp282–314

Vandenberghe, F. (2002) 'Reconstructing humants: A humanist critique of actant-network theory', *Theory, Culture & Society*, 19, no 5/6, pp51–67

Winner, L. (1986) 'Do artifacts have politics?' *The Whale and the Reactor*, University of Chicago Press, Chicago, pp26–38

Wolf, A. T., Natharius, J. A., Danielson, J. J., Ward, B. S. and Pender, J. K. (1999) 'International river basins of the world', *Water Resources Development*, vol 15, no 4, pp387–427

Wouters, P. (1997) 'Present status of international water law', in P. Wouters (ed) *International Water Law*, Kluwer Law International, London

9

The Science–Policy–Stakeholder Interface in Water Management: Lessons Learned and the Challenges Ahead

Geoffrey D. Gooch and Per Stålnacke

Introduction

> Up to 80 to 90 per cent of European research may be wasted because it does not cross the science policy divide. (Phillipe Quevauviller, scientific officer at Directorate-General Research, European Commission)

The preceding chapters in this book have introduced the possibilities and challenges faced during interactions between science, policy-makers and stakeholders, within what we have termed as the science–policy–stakeholder interface (SPSI), and analysed the theoretical frameworks lying behind these interactions, as well as providing examples of how these interactions were practically implemented in the four case basins of the STRIVER research project. We have shown that while the successful interaction and cooperation between research, policy and stakeholders is vital, it is often not straightforward and unproblematic. Within the available literature there are a number of examples drawn upon where efforts made towards constructive and successful cooperation between research, policy-making and stakeholders has not functioned well, and where the interaction was marked by frustration, misunderstandings, disappointment and a lack of substantial progress or positive results. The question, then, is why, and what can we do about it? The chapters in this book have provided a considerable number of examples of how it is possible to improve the SPSI, and in this final chapter we provide some case-basin specific conclusions and recommendations and identify a number of more generic recommendations for improving the SPSI. This chapter concludes by discussing what we see as the major challenges facing the SPSI in water management in the future.

SPSI in the case basins

The basic motivation for the choice of the case basins included in this project was provided in Chapter 1: each one represented a different geographical and climatic context, while at the same time each basin had a number of common characteristics. The common characteristics that they all share is that each river is heavily modified through the development of hydroelectric power (HEP) stations and, in two of the cases (the Tagus and the Tungabhadra), also through intensive irrigation schemes. Two of the rivers are transboundary (the Tagus and the Sesan) and another (the Tungabhadra) flows between two of India's large states. We can therefore say that in some ways we used a 'most similar' choice of cases (heavily modified) and in others we used a 'most dissimilar' choice of cases (geographical and climatic differences). Short descriptions of the case basins are provided in Chapter 1; some details are also provided in the other chapters.

The basis for this book was a scientific research project. The role of science in the SPSI was introduced in Chapter 2 and has been described in varying levels of detail in most of the following chapters. While we noted that there exist significant differences between the views of representatives of different scientific disciplines, the scientific community is still comparatively in agreement about at least some basic ideas. A geographer from Vietnam can talk to and understand a geographer from Spain, and share some common methods and data. A hydrologist or social scientist from different countries can also understand (while they may not agree with) the views of other scientists. Members of the other two groups, policy-makers and stakeholders, however, may contain a more diverse set of participants. As we have pointed out, a politician from Vietnam and his or her counterpart from Cambodia do not always understand one another, and stakeholders from India may well have very different aims and points of view from stakeholders in Spain. We therefore need to say something more about the context within which SPSI works in order to suggest case-specific and general recommendations. In the following sections we will, initially and very briefly, present the SPSI context for each of the case basins based on the previous chapters, then continue with more general comments and conclude with evaluations of the main methods used in the four case areas.

The Tungabhadra River (India)

India has a federal system of governance where water is not under the authority of the union government but a state issue. The two riparian states in the Tungabhadra Basin, Karnataka and Andhra Pradesh, each have their own set of different water governance regimes in place, covering their water policies, laws and legislations, institutions and practices. In this respect the SPSI in the basin shares some of the characteristics of the transboundary rivers, the Sesan and Tagus, as interactions *within* the policy-making group stretch over two different administrative systems (see Chapter 8). The main problems facing water management in the basin were described in Chapters 5 and 6 on water pollution and land and water interaction, respectively. Earlier chapters have also pointed out that there is a wide variety of actors in the basin, and that while civil society organizations are active, there is a clear lack of information-sharing between SPSI actors in the Tungabhadra case area. The proliferation of actors, state and non-state, and the lack of clear lines of communication and information-sharing indicate that the creation of a system to facilitate the SPSI could contribute significantly to further effective and sustainable water management within the basin. This should also include the coordination of water authorities in the two riparian states.

The Sesan River (Vietnam and Cambodia)

As noted in Chapter 8, the interaction of two independent states, with all that involves regarding cooperation between legal systems and institutions, creates special problems for the SPSI. In the case of the Sesan, these interfaces were further complicated by differences in power between the upstream state, Vietnam, and the downstream state, Cambodia, as well as by historical animosity. Water management on the Sesan is also handicapped by a lack of data (especially environmental data), which results in a weaker position for science than in some of the other basins. While efforts are being made to rectify this, lack of reliable data necessitated the use of innovative methods in the basin. In Chapter 7 we described how this lack of data led to the use of a method for determining environmental flow that was dependent on a combination of scientific and local knowledge. Here stakeholders and scientists were shown to be able to work closely together, while policy-makers, due to a lack of funds and competencies at the regional and local levels, often found themselves unable to contribute to knowledge production.

As in the case of the Tungabhadra, insufficient channels of communication between SPSI actors were a major problem. While the Mekong River Commission (MRC) facilitates communication and interaction between national governments through the national MRC commissions, regional and local authorities faced difficulty in gaining information, especially in the case of information-exchange *between* Vietnam and Cambodia at these regional and local levels.

The Tagus River (Spain and Portugal)

While differences exist in the political systems of the two countries, as Spain is a semi-federal state where the regions have decision-making power over a number of significant issues, whereas Portugal is a centralized state with a strong central government, their water management systems are in some ways similar to each other. In this regard, water management on the Tagus differs from that on the Sesan, where water management systems differ both in form and in resources. Both Spain and Portugal are members of the European Union and, as has been pointed out in earlier chapters, are therefore obliged to implement the European Community 2000 Water Framework Directive (WFD), which stipulates an integrated river basin approach to water management. The two countries each have an organization for coordinating the management of the Tagus: the Tejo Basin Council in Portugal and the Tajo Council of Users in Spain. As noted in earlier chapters, both countries have also signed the Albufeira Convention, which regulates the transboundary management of the Tagus. The interaction of policy-makers within the SPSI in the Tagus therefore seems reasonably well developed. Stakeholders were involved in the planning of the Tagus River Basin Plan in Portugal by being provided the opportunity to participate via public discussions. However, methods for improving SPSI in decision-making are not yet in place, and the creation of a stakeholder forum, or river basin committee, for the whole of the Tagus could be one way of improving SPSI interaction.

The Glomma River

Chapter 7, which focuses on the SPSI in environmental flow methodologies and policy for hydropower water release, described the management structures in place for the Glomma River. We pointed out that the Norwegian Ministry of Environment has overall responsibility for water quality management and for the implementation of the WFD, while the Norwegian Ministry of Oil and Energy has overall responsibility for managing water as a resource. Regional authorities at the county level and municipalities are also among the SPSI policy actors. Hence, while there are a number of different administrative levels, they are usually well coordinated, although there may be, as in all countries with this three-level system of administration, problems of communication between levels. Other actors in the Glomma SPSI include the user organizations and NGOs (see Chapters 5 and 7).

Overall reflections on SPSI in the case basins

As is evident from the previous discussion and from the chapters in this book, the situation in different case river basins is quite diverse, just as the SPSI processes in these basins are also distinct. However, despite their various contextual natures, there are certain common features that can be identified in the SPSI processes in the basins and that can be used to formulate a number of general recommendations.

First, there is a growing recognition of IWRM as a framing principle for basin-wide planning and policy action. Irrespective of the actual degree of integration of water management at the basin level and irrespective of the somewhat different meanings that IWRM may have for the different actors in the SPSI, over the last few years increasing importance is being placed on IWRM by the state as well as by non-state actors and stakeholders. This underlines the importance of IWRM as a 'boundary concept' (see Chapter 2), a concept that even though it may have different meanings for different people, nevertheless brings them together and allows a dialogue to take place.

Second, the importance of stakeholder participation is now being increasingly recognized in all the basins. However, statutorily as well as in practice, the process of stakeholder participation leaves a lot to be desired in most of the basins, perhaps with the exception of the Glomma River. While it is usually accepted that stakeholder participation needs to be firmly institutionalized at various stages of water planning, water use, water management and water policy-making (see Chapter 3), this has not always occurred in the case basins. The STRIVER project has provided an innovative approach to this dilemma through the involvement of stakeholders in the process of modelling in the Glomma and Tungabhadra, and scenario-building in all basins, with special focus on the Tagus and Sesan rivers. It has shown that stakeholder participation can lead to better and more relevant formulation of scenarios and easier uptake of the results by the stakeholders.

Pollution as an issue was taken up in two basins: the Tungabhadra and the Glomma. In both cases there was civil society involvement on the issue of pollution. However, since there was little participative infrastructure in place in India, the participation of stakeholders often had to take an adversarial form and there was a series of debates before the Tungabhadra Watchdog Committee was formed. In contrast, in the Glomma Basin, where the participative infrastructure was better developed, stakeholders could come together on a more positive basis, and as they were supported by a consensus on the WFD, could act to bring down pollution levels to safe limits. Similarly, in spite of scientific monitoring protocols, as density and frequency of observations were different in the two basins, a common need was expressed in terms of scientific observations and monitoring.

Environmental flows are important in river basin management since they set the norms for minimum flows to be maintained in rivers. However, this is an issue on which there is great methodological diversity and no clear cut scientific consensus. It is interesting that both in the Glomma Basin as well as the Sesan Basin, a common methodology was developed and applied. The Pressure Impact Multi-Criteria Environmental Flow Analysis (PIMCEFA) method that evolved (see Chapter 7) incorporated inputs from scientists as well non-scientist stakeholders and from formal scientific pools of knowledge, as well as from informal local wisdom. By incorporating local stakeholder opinion, it also ensured that there was a balance between environmental and livelihood needs.

On cross-boundary problems, a major implication was that that dialogue and sustained contact was important. In the case of the Tagus Basin, a convention and an agreement already existed between the two riparian states of Portugal and Spain, and in spite of strains, the constant dialogue between them has allowed them to bring greater cooperation to the table. The WFD has also played an important role in bringing riparian countries together and engaging them in a cooperative dialogue with common objectives. In the case of the Sesan, the basin-wide Mekong Agreement was adopted in 1995 along with the Mekong River Commission. The commission has played an influential role in how waters are managed between the states of the lower Mekong, although significant challenges remain, ensuring that the spirit of the agreement is fully implemented. Some of the riparian states are not yet part of the commission. Nevertheless, there is progress and it is important that apart from allocation issues, environmental issues have also been taken on board by the commission.

The study of land-use and water-use interactions has brought forward a number of issues in the Tungabhadra and the Tagus basins. Here situations are quite diverse. While policy recommendations in the Tungabhadra have generally focused on strengthening agriculture (more specifically, smallholders' agriculture), the situation in the Tagus is generally that of shrinking agricultural space. Naturally, the policy recommendations in both basins diverge. Improvement of water efficiency, along with appropriate land use and innovative technology and institutional options, have a much more central role to play in ensuring sustainable livelihoods and poverty alleviation in the Tungabhadra. In contrast, agriculture forms a much smaller contribution to livelihoods in the Tagus Basin. Interestingly, in both basins, the rising proportion of urban needs, especially in smaller towns, seems to be an important factor in land-use and water-use interactions.

The pairing of basin studies, the so-called 'twinning', has yielded good results. The comparisons bring out both the advantages as well as limitations on exporting knowledge across national and regional boundaries. For example, the value of something like the WFD in highlighting IWRM, as well as in bringing environmental and quality issues into the mainstream of policy-making is an important achievement. However, it is not possible to simply devise an Indian WFD or a South-East Asian WFD along those lines. Vietnam and Cambodia have a long history of conflict and it is not easy to go beyond them to the degree of close cooperation that a WFD type of framework demands. More importantly, such mechanisms should be developed in a 'bottom-up' approach in order to ensure the effective engagement of all stakeholders within legislation reform and to enhance implementation of the adopted agreement. In India, water has been a state subject and there is no true guiding and, to some extent, binding framework at the union or national level. In both of these places, there is a need for a transitional space – first in the spatial sense, in the sense of nested scales, and second, in the sense of moving incrementally, ensuring broader continuities.

Lastly, the STRIVER experience has been important in highlighting the importance of the SPSI processes. It may be emphasized that SPSI is more than a trialogue: while the two poles of science and policy are quite clear cut, the third in terms of stakeholders is more diffuse and diverse. One of the important lessons of the STRIVER project has been that if this trialogue has to take place in a systematic and meaningful manner, then it has to be backed up by policy and laws, and should also have institutional space with access to data and information.

Figure 9.1 Children on the banks of the Sesan River, Cambodia

Source: G. D. Gooch

Recommendations for improving the SPSI in water management

We conclude this chapter, and this book, with a number of general recommendations concerning methods to improve the SPSI in water management. These have been a reccurring theme throughout the book.

Scenarios

In Chapter 4 we suggested that scenarios can be used as a means of both improving policy strategies and of involving stakeholders and the public in that policy-making. Scenarios present a way for the policy-maker or manager to test ideas about possible futures through exercises that can make clearer the probable results of certain courses of action and implementation of particular governance frameworks. They can also be used as a means of improving public and stakeholder participation. If these groups are involved in the formulation and evaluation of scenarios, they can provide insights not readily available for policy-makers, as well as increasing the level of social learning.

Scenarios describe plausible futures; however, the utility of both the process and the products are also vested in the present and can help to identify contemporary problems and issues. Knowledge about key drivers and uncertainties of the future can provide information towards better decisions in the present, and a better understanding of key drivers and trajectories of changes will not only clarify the impact of decisions, but may facilitate active countering of undesirable trajectories of change. The identification and characterization of key uncertainties can enable a more structured approach to risk management. Strategies and decisions can be played out in different futures to secure the most beneficial outcome through the most robust approaches with the least risk. The process can produce new knowledge that will not only benefit resource managers and decision-makers, but empower stakeholders. This extends from politicians and policy-makers through to government officials, the private sector and civil society. Knowledge generated and sourced through structured participative research processes can increase understanding both through participating in the process and by accessing appropriate forms of communication. The identification, description and ranking of key drivers and uncertainties can be translated to potential implications. Scenarios have been the focus of increased interest and initiatives, such as those published by the European Environmental Agency on the Environmental Scenarios Information portal (http://scenarios.ew.eea.europa.eu/), and have contributed to developments in this field; yet the use of scenarios as a means of improving stakeholder participation and the science–policy interface still needs to be developed.

In Chapter 4 we demonstrated how scenarios can be used as a means of two-way communication with stakeholders and civil society. The stakeholder groups formulated future possible developments for the project team, which then used them in their analyses. The meeting participants identified who they saw as the main actors and problems. These were then incorporated, together with the results of the analyses of actor networks, communication and legal aspects, within draft scenarios, which were discussed at a second stakeholder workshop. Following this, the scenarios were revised once again and finally presented at a third stakeholder workshop. Through this process, stakeholder perspectives, governance analyses and the results of scientific research in the river basins were combined. The stakeholder workshops also helped to identify central aspects of the project, such as the actor networks. Finally, the development of scenarios highlighted particular aspects of the legal regime that may need to be altered and those aspects of the existing regimes that obstruct, affect or are likely to lead to particular scenarios, or that have an effect on the extent to which particular stakeholder groups can influence resource management.

The use of scenarios showed that they can be a useful tool in involving stakeholders and improving the legitimacy and quality of policy strategies. As the meetings were held on both sides of the Tagus and Sesan rivers, stakeholders were able to formulate national perspectives, which were then combined at meetings with all participants. In the case of the Sesan River, this meeting, held at the MRC in Vientiane in December 2008, was in itself an achievement – a problem that has complicated the formulation of policies for the sustainable use of the Sesan River has been a relative lack of communication between the Vietnamese and Cambodian stakeholders, especially at the provincial level.

The major challenges to the use of scenarios are as follows:

- The process is time consuming, and in order to succeed it is necessary to organize at least three meetings: one to introduce the idea of scenarios, which can be difficult to grasp initially; a second meeting to discuss the first draft versions; and a third to finalize the scenarios.
- The selection of participants for the meetings is a central issue; it is at the same time necessary to create a group that can engage in creative discussions without excluding variant perspectives. This is important: while diametrically opposing perspectives may make the process more difficult, they are necessary if the specific function of scenarios, that of formulating possible futures (not necessarily probable), is to be achieved.
- In rural areas, especially in developing countries, the logistics of gathering people from a wider geographical area can be substantial. In the case of the Sesan River, for example, it proved difficult to arrange a meeting at the local level in the river basin because of demanding communications.
- The choice of factors upon which to base the scenarios is difficult; simplification is necessary and may resolve the complicated interdependency of issues in sustainable water management.
- As with all processes that involve stakeholder participation, it is necessary to define realistic outcomes at the very beginning of the process. The results of an exercise such as this with scenarios can increase the understanding of the participants for common problems, can raise their awareness and knowledge, and can increase their capacity to define their own futures. However, it is necessary to make clear that the impact upon policy made by local, regional and national authorities is completely dependent on the willingness of authorities to take on the results of the process.

Despite these reservations, however, the use of stakeholder-formulated scenarios, especially in the Sesan and Tagus basins, has demonstrated that the method can be a useful contribution to the development of sustainable policies for water management.

SPSI, tools, data and information

Scientific tools and models have an important role to play in water management and its importance has increased during the last decade. For example, in Europe, an increased demand for scientific tools in connection to the implementation of the WFD can be observed. More specifically, accurate quantification of source emissions, river pollution loads and their impact upon mitigation measures, etc. are major challenges facing the research community as managers and policy-makers increasingly rely on outputs from river basin models when evaluating environmental changes and management actions. In a science–policy context, this relates to both the selection of appropriate tools and ensuring that the management scenarios and mitigation measures are actually relevant.

In relation to this, it has also been pointed out a number of times that one of the major challenges facing the SPSI in water management is the production, management and communication of data and information (see, for example, Chapter 8).

We therefore provide here some short examples of the importance of scientific tools, data and information for an efficient SPSI process. The focus is on our experience with modelling nutrients, although some brief discussions about other applied models are also provided.

As discussed in more detail in Chapter 5, we used the same river basin modelling tool in both the Glomma (sub-basins Hunnselva and Lena) and in the Tungabhadra river basins.

Here, we will not present specific modelling results and management recommendations for the Glomma and Tungabhadra river basins; these can be found in Barkved et al (2008, 2009), Lo Porto et al (2008) and Grizzetti et al (2008).

It was found that involving stakeholders in a model exercise, besides ensuring local participation and the relevance of practical management, also increases the likelihood of incorporating local knowledge and understanding of the natural system. In such a process it is important to identify and understand the values and motives of a wide range of stakeholders. Moreover, the joint modelling work involving scientists, water managers and stakeholders was also of practical management value. More specifically, the modelling results of the loading impacts upon mitigation measures in Hunnselva were, in fact, included in the river basin management plan (currently open to public hearing in connection with the implementation of the WFD in Norway; see www.vannportalen.no/enkel.aspx?m=40354).

The overall key message in pollution modelling with stakeholder involvement is that knowledge about the local conditions is an important asset in order to ensure reliable results. The experiences also showed that stakeholder involvement at different phases of the modelling process, such as model input preparation, scenario-building and discussions of modelling outcome, play a key role. Similar experiences were had in connection with the development of the environmental flow method (see Chapter 7), where the main information input came from the local expertise (such as local fishermen along the rivers) rather than from environmental data bases. The identification of actors in the actor network method (see Chapter 8) was also based on the same approach (e.g. by visits to study sites and interviews with local people and managers).

One common feature and clear advantage with all the applied and tested tools was that they promoted dialogue and integration between the different actors of SPSI and IWRM.

However, it was also recognized that there was a lack of data and information to effectively apply the models and methods. For example, in the pollution modelling work, much of the required data were not available, non-existent or not available at the required scale and resolution necessary for model calibration and validation. At the same time, data were rarely owned by the same institution. This problem was most visible in the Indian case; but even for the more 'data-rich' basins in Norway, this was also a challenge (see Chapter 5). Nonetheless, the positive side effect of the interactive process of data acquisition was the increased awareness about the data gaps amongst the managers and data-owning participants and, consequently, the limitations that this can impose on the modelling outcome. Data and information availability was also found to be a hindrance in relation to scenario development (see Chapter 4). This could be seen during the initial stage of scenario development when, for example, describing the physical conditions in the basin. Moreover, the development of the environmental flow method (see Chapter 7) was, in fact, due to a lack of data (such as detailed river profiles and ecological data) that hindered the application of more data-demanding methods. The environmental flow method therefore heavily relied on the interaction of scientists and local expertise, rather than on traditional data inputs.

In order to optimize resource use and to maximize management benefits, we recommend that data collection and monitoring programmes are discussed between water managers and the research community. Without doubt, data and information are crucial for objective river basin system and related scientific understanding, and thus also for efficient management (see also Chapter 8) and communication (see the following sub-section).

SPSI and transboundary rivers

As noted in Chapter 8, conflicts over water and the demands placed on water governance increase in transboundary settings – in these contexts representatives of different national and sectoral interests from different countries must try to cooperate. In the national contexts, water laws and systems of administration are mostly unified (although competition between different agencies can and often does exist). In international rivers there is, with a few exceptions, no unitary authority that can force actors in the water sector to comply with laws and agreements. Therefore, while water management regimes in transboundary rivers are especially dependent on efficient legal systems and communication, they often do not exist. This makes cooperation between the different organizations and institutions, such as governmental agencies and departments, more complicated. At the same time, NGOs and other stakeholder groups may find it difficult to create efficient means of communication and influence with these diverse groups of policy-makers and managers (Gooch, 2008). Communication is a central issue in transboundary water regimes and takes place in a number of networks, which consist of both official and unofficial actors. These networks include government officials, NGOs, village elders, representatives of development banks, etc., as well as the infrastructures within the river basin, such as hydroelectric power plant dams, and should be viewed as actor networks and as networks of influences (Gooch, 2006). Within these networks, the different worldviews and understanding of the actors, and groups of actors, strongly influence their treatment of information and knowledge. The scientific information provided by a project such as STRIVER may be accepted or rejected according to these mind frames (Gooch, 2004).

A key aspect of the work related to water governance was to develop a robust set of indicators for assessing governance within the context of transboundary IWRM. The entry point for such work was the recognition that, while considerable research has sought to measure governance in general, few initiatives have sought to tailor such initiatives to the specific context of (transboundary) IWRM. Having reviewed existing theory related to governance, IWRM and indicator analysis, STRIVER developed a set of indicator questions to examine the extent to which key aspects of *good* governance and IWRM have been embedded within the applicable laws (international, regional and national) and the degree to which such laws have been implemented in practice. Collaboration with stakeholders, through the workshops and subsequent targeted interviews, constituted an essential element in ascertaining the extent to which the applicable laws had been implemented. Stakeholder workshops also provide an important means for 'validating' the research developed by the STRIVER research team. Ultimately, the governance research concluded that IWRM implementation is heavily reliant on the existence of a broader governance framework that supports access to information and justice, as well as stakeholder participation in decision-making (Rieu-Clarke and Allan, 2008).

Stakeholder involvement in SPSI and water management

It has been stressed that stakeholder participation and analysis is one of the most critical elements for the practical implementation of IWRM within the SPSI. Moreover, a central element of achieving good governance is the need to ensure effective *participation* in decision-making procedures, as was clearly addressed in Chapter 8.

Stakeholder participation as a method to improve the SPSI was given an important place in the project, as demonstrated through the series of stakeholder workshops that were conducted in all the four case basins. In all, 12 stakeholder workshops were conducted (three within each basin), in addition to a range of targeted discussions with key stakeholders. During the first year, the workshops provided a platform for reviewing the initial stakeholder analyses and identifying the key stakeholders; introducing stakeholders to the project objectives and mapping stakeholder expectations, interests and problems; and fostering synergies with ongoing activities within each of the basins. During the second year, stakeholder workshops were used to steer research objectives and activities, including developing policy scenarios. The final-year workshops were essential as a tool for collaboratively reviewing project outputs with stakeholders and the STRIVER team, as well as identifying avenues for further exploitation of STRIVER results.

The workshops not only helped to integrate the various perspectives of stakeholders from different sectors, but also from different user groups with varied and conflicting interests. Experience showed that there was a relatively strong willingness among stakeholders to embrace the IWRM process irrespective of country, sector and/or occupational background, although the modalities remained fuzzy. The group dynamics observed at the workshops proved that it was possible to bring stakeholders who shared these water bodies together for constructive dialogue, although the political, cultural and institutional context within each of the basins had a major impact upon participation. Research project-focused stakeholder workshops proved to be a useful tool for enabling soft negotiations on transboundary management of water resources and for identifying opportunities for resolving other water-use conflicts. It was also noted that projects such as STRIVER could play a 'neutral role' in moderating the stakeholder workshops and motivating stakeholders with conflicting interests by presenting research findings that were perceived to possess a strong air of legitimacy. The stakeholder workshops also played an important role in offering insights on IWRM practice from other basins around the world and thus promoting awareness, as well as, to some extent, capacity-building. Ultimately, the stakeholder workshops helped in fostering linkages between the STRIVER researchers, managers, end users and policy-makers, and at the same time improved acceptance of project outcomes.

Reflections

Compared to many other research and development (R&D) projects, twinning projects such as STRIVER have the flexibility to interact with water managers and policy-makers. However, this requires that scientists take the role of 'scientific ambassadors'. In addition, it also demands more time and resources in terms of logistics required for travelling, meetings and discussion forums. Managers and policy-makers may initially be sceptical about spending their valuable time with researchers, as was observed during the initial stages of STRIVER. In the course of time, however, STRIVER managed to establish trust and confidence amongst the stakeholders. At the same time, during the last decade, scientific publication has become an ever more important criterion for the monitoring of research and researchers (e.g. through national research councils) of the degree of scientific success at both academic institutions and research institutes. Thus, for researchers, stakeholder interactions and practical involvement in the SPSI, as recommended in this book, can be at the expense of scientific publications. The career of scientists largely depends on their scientific output (in journals) and only to a much lesser extent on whether they are involved in policy advice. Thus, in terms of career prospects, there is little incentive for scientists to be actively involved in policy advice and stakeholder interactions.

If results from research projects have to reach managers and policy-makers, it is therefore necessary to commit the resources and establish formal links between research, policy and stakeholders (from local-level to high-level water managers and policy-makers). Such links could constitute a win–win situation for both researchers and managers. They could also help in developing a better understanding of the research problems to be addressed, and in jointly developing appropriate tools, scenarios and policy guidelines. We have also experienced – during the field trips and workshops – that stakeholders were interested in capacity-building which could be also formalized in the projects. In more practical terms, it is recommended that national funding organizations and multinational agencies, such as Directorate-General Research at the European Commission, should have more emphasis on twinning between scientific community and stakeholders and should involve at least four to five years of project duration. The experiences and examples given in this book have clearly shown that this is the only viable way forward. Research projects such as STRIVER can facilitate identifying and addressing challenges related to SPSI and IWRM in a given river basin. Additionally, such projects can contribute towards improving stakeholder integration and participation. In STRIVER, the experience and knowledge of stakeholders complemented discussions about IWRM grounded in science-based results. Stakeholders in the basins demonstrated a special interest in the scientific tools and results developed and applied in the project (e.g. on pollution modelling and environmental flow), and the scenarios were developed jointly by the scientists and the stakeholders during three stakeholder workshops in each of the four case basins.

Conclusions: SPSI in water management

The chapters in this book, authored by a large group of researchers coming from many different disciplines, have attempted to unravel the science–policy–stakeholder interface and, at the same time, provide recommendations on how improvements in the SPSI will help water management become more efficient and sustainable. The most important conclusions coming from the work reported in this publication are:

- Stakeholder participation is one of the most critical elements for practical IWRM implementation and a central aspect of the SPSI. Participation helps not only in integrating the various perspectives of stakeholders from different sectors, but also of different user groups with varied and conflicting interests. Stakeholder participation can also provide valuable insights into water management issues and provide information not otherwise available, especially in data-poor parts of the world (see especially Chapter 5).
- Research projects such as STRIVER can act as independent facilitators and provide a neutral platform for SPSI dialogue, which ultimately can facilitate the IWRM process. If such projects can gain the trust of all actors in the SPSI, then they can help to improve communication and information exchange.
- Transboundary rivers present a special challenge to the SPSI as they involve the interaction of more diverse actor groups than are usually found in national contexts. Project workshops that bring together actor groups from

different countries, preferably held in a third 'neutral' county, can help to create trust and develop future contacts.
- The development and show cases of various water management 'tools', such as environmental flow, pollution models, water pricing, actor network analyses and scenarios, were of significant interest for water managers. One common feature with all the applied and tested tools was that they promoted dialogue and integration between different SPSI actors.

References

Barkved, L. J., Deelstra, J., Grizzetti, B. and Bouroui, F. (2008) *Modelling Nutrients in the Glomma River Basin: Scenarios and Management Recommendations*, STRIVER Policy Brief No 11, http://kvina.niva.no/striver/Portals/0/documents/STRIVER_PB11.pdf, last accessed March 2010

Barkved, L. J., Fazi, S. and Lo Porto, A. (eds) (2009) *Scientific Report on Pollution Source Assessment, Including Source Apportionment Results, and Pollution Prevention Measures*, STRIVER Report No D7.1, Part 2, http://kvina.niva.no/striver/Portals/0/documents/STRIVER_D7_1_Part2.pdf, last accessed March 2010

Barton, D. N., Raju, K. V. and McNeill, D. (2008) *Water Valuation and Pricing – When Are They Useful in Water Management?*, STRIVER Policy Brief No 3, http://kvina.niva.no/striver/Portals/0/documents/STRIVER_PB3.pdf, last accessed 2010

Gooch, G. D. (2004) 'The communication of scientific information in institutional contexts: The specific case of transboundary water management in Europe', in J. G. Timmerman and S. Langaas (eds) *Environmental Information in European Transboundary Water Management*, IWA Publishing, London

Gooch, G. D. (2006) 'Actor network theory as a tool for analyses of multi-level water governance', Paper presented to the International Workshop on Governance and the Global Water System: Institutions, Actors, Scales of Water Management Facing the Challenges of Global Change, Global Water System Project, Bonn, Germany, 20–23 June 2006

Gooch, G. D. (2008) *The Use and Communication of Information in Transboundary Water Regimes: The Case of the Tagus and Sesan Rivers*, STRIVER Report No D6.3, Submitted to the EC, http://kvina.niva.no/striver/Portals/0/documents/STRIVER_D6_2.pdf, last accessed March 2010

Grizzetti, B., Lo Porto, A., Barkved, L. J., Bouraoui, F., Deelstra, J. and Joy, K. (2008) *Modelling Water Pollution with Stakeholders Involvement: The Twinned Experience of Glomma (Norway) and Tungabhadra (India) River Basins*, STRIVER Policy Brief No 10, http://kvina.niva.no/striver/Portals/0/documents/STRIVER_PB10.pdf

Lo Porto, A., Barkved, L. J. and Gosain K. A. (2008) *Modelling Water and Nutrients Balance in Tungabhadra River Basin: Scenario Analysis and Management Recommendations*, STRIVER Policy Brief No 7, http://kvina.niva.no/striver/Portals/0/documents/STRIVER_PB7.pdf, last accessed March 2010

Rieu-Clarke, A. and Allan, A. (eds) (2008) *Role of Water Law: Assessing Governance in the Context of IWRM: An Analysis of Commitment and Implementation in the Tagus and Sesan River Basins*, STRIVER Report No D6.3, http://kvina.niva.no/striver/Portals/0/documents/STRIVER_D6_3.pdf, last accessed March 2010

Index